SCIENTIFIC INFERENCE

Providing the knowledge and practical experience to begin analysing scientific data, this book is ideal for physical sciences students wishing to improve their data handling skills.

The book focuses on explaining and developing the practice and understanding of basic statistical analysis, concentrating on a few core ideas, such as the visual display of information, modelling using the likelihood function, and simulating random data.

Key concepts are developed through a combination of graphical explanations, worked examples, example computer code and case studies using real data. Students will develop an understanding of the ideas behind statistical methods and gain experience in applying them in practice. Further resources are available at www.cambridge.org/9781107607590, including data files for the case studies so students can practice analysing data, and exercises to test students' understanding.

SIMON VAUGHAN is a Reader in the Department of Physics and Astronomy, University of Leicester, where he has developed and runs a highly regarded course for final year physics students on the subject of statistics and data analysis.

SCIENTIFIC INFERENCE

Learning from data

SIMON VAUGHAN
University of Leicester

CAMBRIDGE
UNIVERSITY PRESS

Shaftesbury Road, Cambridge CB2 8EA, United Kingdom

One Liberty Plaza, 20th Floor, New York, NY 10006, USA

477 Williamstown Road, Port Melbourne, VIC 3207, Australia

314–321, 3rd Floor, Plot 3, Splendor Forum, Jasola District Centre, New Delhi – 110025, India

103 Penang Road, #05–06/07, Visioncrest Commercial, Singapore 238467

Cambridge University Press is part of Cambridge University Press & Assessment, a department of the University of Cambridge.

We share the University's mission to contribute to society through the pursuit of education, learning and research at the highest international levels of excellence.

www.cambridge.org
Information on this title: www.cambridge.org/9781107024823

First published 2013

A catalogue record for this publication is available from the British Library

Library of Congress Cataloging-in-Publication data
Vaughan, Simon, 1976– author.
Scientific inference : learning from data / Simon Vaughan.
pages cm
Includes bibliographical references and index.
ISBN 978-1-107-02482-3 (hardback) – ISBN 978-1-107-60759-0 (paperback)
1. Mathematical statistics – Textbooks. I. Title.
QA276.V34 2013
519.5 – dc23 2013021427

ISBN 978-1-107-02482-3 Hardback
ISBN 978-1-107-60759-0 Paperback

For my family

Contents

For the student

Science is not about certainty, it is about dealing rigorously with uncertainty. The tools for this are statistical. Statistics and data analysis are therefore an essential part of the scientific method and modern scientific practice, yet most students of physical science get little explicit training in statistical practice beyond basic error handling. The aim of this book is to provide the student with both the knowledge and the practical experience to begin analysing new scientific data, to allow progress to more advanced methods and to gain a more statistically literate approach to interpreting the constant flow of data provided by modern life.

More specifically, if you work through the book you should be able to accomplish the following.

- Explain aspects of the scientific method, types of logical reasoning and data analysis, and be able to critically analyse statistical and scientific arguments.
- Calculate and interpret common quantitative and graphical statistical summaries.
- Use and interpret the results of common statistical tests for difference and association, and straight line fitting.
- Use the calculus of probability to manipulate basic probability functions.
- Apply and interpret model fitting, using e.g. least squares, maximum likelihood.
- Evaluate and interpret confidence intervals and significance tests.

Students have asked me whether this is a book about statistics or data analysis or statistical computing. My answer is that they are so closely connected it is difficult to untangle them, and so this book covers areas of all three.

The skills and arguments discussed in the book are highly transferable: statistical presentations of data are used throughout science, business, medicine, politics and the news media. An awareness of the basic methods involved will better enable you to use and critically analyse such presentations – this is sometimes called *statistical literacy*.

In order to understand the book, you need to be familiar with the mathematical methods usually taught in the first year of a physics, engineering or chemistry degree (differential and integral calculus, basic matrix algebra), but this book is designed so that the probability and statistics content is entirely self-contained.

For the instructor

This book was written because I could not find a suitable textbook to use as the basis of an undergraduate course on scientific inference, statistics and data analysis. Although there are good books on different aspects of introductory statistics, those intended for physicists seem to target a post-graduate audience and cover either too much material or too much detail for an undergraduate-level first course. By contrast, the 'Intro to stats' books aimed at a broader audience (e.g. biologists, social scientists, medics) tend to cover topics that are not so directly applicable for physical scientists. And the books aimed at mathematics students are usually written in a style that is inaccessible to most physics students, or in a recipe-book style (aimed at science students) that provides ready-made solutions to common problems but develops little understanding along the way.

This book is different. It focuses on explaining and developing the *practice* and *understanding* of basic statistical analysis, concentrating on a few core ideas that underpin statistical and data analysis, such as the visual display of information, modelling using the likelihood function, and simulating random data. Key concepts are developed using several approaches: verbal exposition in the main text, graphical explanations, case studies drawn from some of history's great physics experiments, and example computer code to perform the necessary calculations.[1] The result is that, after following all these approaches, the student should both understand the ideas behind statistical methods and have experience in applying them in practice.

The book is intended for use as a textbook for an introductory course on data analysis and statistics (with a bias towards students in physics) or as self-study companion for professionals and graduate students. The book assumes familiarity with calculus and linear algebra, but no previous exposure to probability or statistics

[1] These are based on R, a freely available software package for data analysis and statistics and used in many statistics textbooks.

is assumed. It is suitable for a wide range of undergraduate and postgraduate science students.

The book has been designed with several special features to improve its value and effectiveness with students:

- several complete data analysis case studies using real data from some of history's great experiments
- 'example boxes' – approximately 20 boxes throughout the text that give specific, worked examples for concepts as they are discussed
- 'computer practice boxes' – approximately 90 boxes throughout the text that give working R code to perform the calculations discussed in the text or produce the plots shown
- graphical explanations of important concepts
- appendices that provide technical details supplementary to the main text
- a well-populated glossary of terms and list of notational conventions.

The emphasis on a few core ideas and their practical applications means that some subjects usually covered in introductory statistics texts are given little or no treatment here. Rigorous mathematical proofs are not covered – the interested reader can easily consult any good reference work on probability theory or mathematical statistics to check these. In addition, we do not cover some topics of 'classical' statistics that are dealt with in other introductory works. These topics include

- more advanced distribution functions (beta, gamma, multinomial, ...)
- ANOVA and the generalised linear model
- characteristic functions and the theory of moments
- decision and information theories
- non-parametric tests
- experimental design
- time series analysis
- multivariate analysis (principal components, clustering, ...)
- survival analysis
- spatial data analysis.

Upon completion of this book the student should be in a much better position to understand any of these topics from any number of more advanced or comprehensive texts.

Perhaps the 'elephant in the room' question is: what about Bayesian methods? Unfortunately, owing to practical limitations there was not room to include full chapters developing Bayesian methods. I hope I have designed the book in such a way that it is not wholly frequentist or Bayesian. The emphasis on model fitting using the likelihood function (Chapter 6) could be seen as the first step towards a Bayesian analysis (i.e. implicitly using flat priors and working towards the posterior mode). Fortunately, there are many good books on Bayesian data analysis that can then be used to develop Bayesian ideas explicitly. I would recommend Gelman *et al.* (2003) generally and Sivia and Skilling (2006) or Gregory (2005) for physicists in particular. Albert (2007) also gives a nice 'learn as you compute' introduction to Bayesian methods using R.

1

Science and statistical data analysis

> It is remarkable that a science which began with the consideration of games of chance should have become the most important object of human knowledge.
>
> Pierre-Simon Laplace (1812)
> *Théorie Analytique des Probabilités*

Why should a scientist bother with statistics? Because science is about dealing rigorously with uncertainty, and the tools to accomplish this are statistical. Statistics and data analysis are an indispensable part of modern science.

In scientific work we look for relationships between phenomena, and try to uncover the underlying patterns or laws. But science is not just an 'armchair' activity where we can make progress by pure thought. Our ideas about the workings of the world must somehow be connected to what actually goes on in the world. Scientists perform experiments and make observations to look for new connections, test ideas, estimate quantities or identify qualities of phenomena. However, experimental data are never perfect. Statistical data analysis is the set of tools that helps scientists handle the limitations and uncertainties that always come with data. The purpose of statistical data analysis is *insight* not just *numbers*. (That's why the book is called *Scientific Inference* and not something more like *Statistics for Physics*.)

1.1 Scientific method

Broadly speaking, science is the investigation of the physical world and its phenomena by experimentation. There are different schools of thought about the philosophy of science and the scientific method, but there are some elements that almost everyone agrees are components of the scientific method.

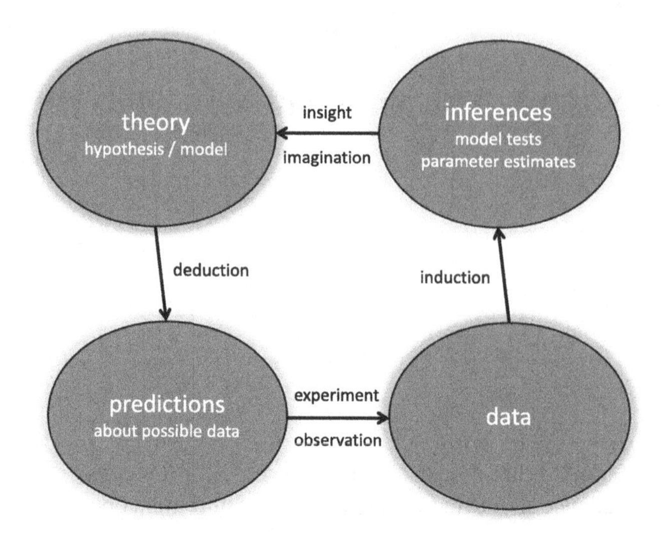

Figure 1.1 A cartoon of a simplified model of the scientific method.

Hypothesis A hypothesis or model is an explanation of a phenomenon in terms of others (usually written in terms of relations or equations), or the suggestion of a connection between phenomena.

Prediction A useful hypothesis will allow predictions to be made about the outcome of experiments or observations.

Observation The collection of experimental data in order to investigate a phenomenon.

Inference A comparison between predictions and observations that allows us to learn about the hypothesis or model.

What distinguishes science from other disciplines is the insistence that ideas be tested against what actually happens in Nature. In particular, hypotheses must make predictions that can be tested against observations. Observations that match closely the predictions of a hypothesis are considered as evidence in support of the hypothesis, but observations that differ significantly from the predictions count as evidence against the hypothesis. If a hypothesis makes no predictions about possible observations, how can we learn about it through observation?

Figure 1.1 gives a summary of a simplified scientific method. Models and hypotheses[1] can be used to make predictions about what we can observe.

[1] The terms 'hypothesis', 'model' and 'theory' have slightly different meanings but are often used interchangeably in casual discussions. A *theory* is usually a reasonably comprehensive, abstract framework (of definitions, assumptions and relations or equations) for describing generally a set of phenomena, that has been tested and found at least some degree of acceptance. Examples of scientific theories are classical mechanics, thermodynamics, germ theory, kinetic theory of gases, plate tectonics etc. A *model* is usually more specific. It might be the application of a theory to a particular situation, e.g. a classical mechanics model of the orbit of Jupiter. Some

Hypotheses may come from some more general theory, or may be more ad hoc, based on intuition or guesswork about the way some phenomenon might work. Experiments or observations of the phenomenon can be made, and the results compared with the predictions of the hypothesis. This comparison allows one to *test* the model and/or *estimate* any unknown parameters. Any mismatch between data and model predictions, or other unpredicted findings in the data, may suggest ways to revise or change the model. This process of learning about hypotheses from data is scientific inference. One may enter the cycle at any point: by proposing a model, making predictions from an existing model, collecting data on some phenomenon or using data to test a model or estimate some of its parameters. In many areas of modern science, the different aspects have become so specialised that few, if any, researchers practice all of these activities (from theory to experiment and back), but all scientists need an appreciation of the other steps in order to understand the 'big picture'. This book focuses on the induction/inference part of the chain.

1.2 Inference

The process of drawing conclusions based on what is already known is called *inference*. There are two types of reasoning process used in inference: deductive and non-deductive.

1.2.1 Deductive reasoning (from general to specific)

The first kind of reasoning is *deductive reasoning*. This starts with premises and follows the rules of logic to arrive at conclusions. The conclusions are therefore true as long as the premises are true. Philosophers say the premises entail the conclusion. Mathematics is based on deductive reasoning: we start from axioms, follow the rules of logic and arrive at theorems. (Theorems should be distinguished from theories – the former are the product of deductive reasoning; the latter are not.) For example, the two propositions 'A is true implies B is true' and 'A is true' together imply 'B is true'. This type of argument is a simple deduction known as a syllogism, which comprises a major premise and a minor premise; together they imply a conclusion:

$$\begin{aligned} \text{Major premise} &: A \Rightarrow B \text{ (read: } A \text{ is true implies } B \text{ is true)} \\ \text{Minor premise} &: A \text{ (read: } A \text{ is true)} \\ \text{Conclusion} &: B \text{ (read: } B \text{ is true).} \end{aligned}$$

Deductive reasoning leads to conclusions, or theorems, that are inescapable given the axioms. One can then use the axioms and theorems together to deduce more

authors go on to distinguish *hypotheses* as models, and their parameters, which may be speculative, as they are used in statistical inference. For now we have no need to distinguish between models and hypotheses.

theorems, and so on. A theorem[2] is something like '$A \Rightarrow B$', which simply says that the truth value of A is transferred to B, but it does not, in and of itself, assert that A or B are true. If we happen to know that A is indeed true, the theorem tells us that B must also be true. The box gives a simple proof that there is no largest prime number, a purely deductive argument that leads to an ineluctable conclusion.

Box 1.1
Deduction example – proof of no largest prime number

- Suppose there is a largest prime number; call this p_N, the Nth prime.
- Make a list of each and every prime number: $p_1 = 2$, $p_2 = 3$, $p_3 = 5$, until p_N.
- Now form a new number q from the product of the N primes in the list, and add one:

$$q = 1 + \prod_{i=1}^{N} p_i = 1 + (p_1 \times p_2 \times p_3 \times \cdots \times p_N) \tag{1.1}$$

which is either prime or it is not.
- This new number q is larger than every prime in the list, but it is not divisible by any prime in the list – it always leaves a remainder of one.
- This means q is prime since it has no prime factors (the fundamental theorem of arithmetic says that any integer larger than 1 has a unique prime factorisation).
- But this is a contradiction. We have found a prime number q that is larger than every number in our list, in contradiction with our definition of p_N. Therefore our original assumption – that there is a largest prime, p_N – must be false.

Deduction involves reasoning from the general to the specific. If a general principle is true, we can conclude that any particular cases satisfying the general principle are true. For example:

Major premise : All monkeys like bananas
Minor premise : Zippy is a monkey
Conclusion : Zippy likes bananas.

The conclusion is unavoidable given the premises. (This type of argument is given the technical name *modus ponens* by philosophers of logic.) If some theory is true we can predict that its consequences must also be true. This applies to probabilistic as well as deterministic theories. Later on we consider flipping coins, rolling dice, and other random events. Although we cannot precisely predict the outcome of

[2] It is worth noting here that the logical implication used above, e.g. $B \Rightarrow A$, does not mean that A can be derived from B, but only that if B is true then A must also be true, or that the propositions 'B is true' and 'B and A are both true' must have the same truth value (both true, or both false).

individual events (they are random!), we can derive frequencies for the various outcomes in repeated events.

1.2.2 Inductive reasoning (from specific to general)

Inductive reasoning is a type of non-deductive reasoning. Induction is often said to describe arguments from special cases to general ones, or from effects to causes. For example, if we observe that the Sun has risen every day for many days, we can inductively reason that it will continue to do so. We cannot directly deduce that the Sun will rise tomorrow (there is no logical contradiction implied if it does not).

The basic point about the limited power of our inferences about the real world (i.e. our inductive reasoning) was made most forcefully by the Scottish philosopher David Hume (1711–1776), and is now known as the problem of induction. The philosopher and mathematician Bertrand Russell furnished us with a memorable example in his book *The Problems of Philosophy* (Russell, 1997, ch. 4):

> imagine a chicken that gets fed by the farmer every day and so, quite understandably, imagines that this will always be the case . . . until the farmer wrings its neck! The chicken never expected that to happen; how could it? – given it had no experience of such an event and the uniformity of its previous experience had been so great as to lead it to assume the pattern it had always observed (chicken gets fed every day) was universally true. But the chicken was wrong.[3]

You can see that inductive reasoning does not have the same power as deductive reasoning: a conclusion arrived at by deductive reasoning is necessarily true if the premises are true, whereas a conclusion arrived at by inductive reasoning is not *necessarily* true, it is based on incomplete information. We cannot deduce (prove) that the Sun will rise tomorrow, but nevertheless we do have confidence that it will. We might say that deductive reasoning concerns statements that are either true or false, whereas inductive reasoning concerns statements whose truth value is unknown, about which we are better to speak in terms of 'degree of belief' or 'confidence'. Let's see an example:

Major premise : All monkeys we have studied like grapes
Minor premise : Zippy is a monkey
Conclusion : Zippy likes grapes.

The conclusion is not unavoidable, other conclusions are allowed. There is no logical contradiction in concluding

Conclusion : Zippy does not like grapes.

[3] By permission of Oxford University Press.

But the premises do give us some information. It seems plausible, even probable, that Zippy likes grapes.

1.2.3 Abductive reasoning (inference to the best explanation)

There is another kind of non-deductive inference, called *abduction*, or *inference to the best explanation*. For our purposes, it does not matter whether abduction is a particular type of induction, or another kind of non-deductive inference alongside induction. Let's go straight to an example:

$$\begin{aligned} \text{Premise} &: \text{Nelly likes bananas} \\ \text{Premise} &: \text{The banana left near to Nelly has been eaten} \\ \text{Conclusion} &: \text{Nelly ate the banana.} \end{aligned}$$

Again the conclusion is not unavoidable, other conclusions are valid. Perhaps someone else ate the banana. But the original conclusion seems to be in some sense the simplest of those allowed. This kind of reasoning, from observed data to an explanation, is used all the time in science.

Induction and abduction are closely related. When we make an inductive inference from the limited observed data ('the monkeys in our sample like grapes') to unobserved data ('Zippy likes grapes') it is as if we implicitly passed through a theory ('all monkeys like grapes') and then deduced the conclusion from this.

1.3 Scientific inference

Scientific work employs all the above forms of reasoning. We use deductive reasoning to go from general theories to specific predictions about the data we could observe, and non-deductive reasoning to go from our limited data to general conclusions about unobserved cases or theories.

Imagine A is the theory of classical mechanics and B is the predicted path of a rocket deduced from the theory and the details of the launch. Now, we make some observations and find the rocket did indeed follow the predicted path B (as well as we can determine). Can we conclude that A is true? We may infer A, but not deductively. Other conclusions are possible. In fact, the observational confirmation of one prediction (or even a thousand) does not *prove* the theory in the same sense as a deductive proof. A different theory may make indistinguishable predictions in all of the cases considered to date, but differ in its predictions for other (e.g. future) observations.

Experimental and observational science is all about inductive reasoning, going from a finite number of observations or results to a general conclusion about unobserved cases (induction), or a theory that explains them (abduction). In recent years, there has been a lot of interest in showing that inductive reasoning can be formalised in a manner similar to deductive reasoning, so long as one allows for the uncertainty in the data and therefore in the conclusions (Jeffreys, 1961; Jaynes, 2003).

You might still have reservations about the need for statistical reasoning. After all, the great experimental physicist Ernest Rutherford is supposed to have said

> If your experiment needs statistics, you ought to have done a better experiment![4]

Rutherford probably didn't say this, or didn't mean for it to be taken at face value. Nevertheless, statistician Bradley Efron, about a hundred years later, contrasted this simplistic view with the challenges of modern science (Efron, 2005):

> Rutherford lived in a rich man's world of scientific experimentation, where nature generously provided boatloads of data, enough for the law of large numbers to squelch any noise. Nature has gotten more tight-fisted with modern physicists. They are asking harder questions, ones where the data is thin on the ground, and where efficient inference becomes a necessity. In short, they have started playing in our ball park.

But it is not just scientists who use (or should use) statistical data analysis. Any time you have to draw conclusions from data you will make use of these skills. This is true for particle physics as well as journalism, and whether the data form part of your research or come from a medical test you were given you need to be able to understand and interpret them properly, making inferences using methods built on the same basic principles.

1.4 Data analysis in a nutshell

The analysis of data[5] can be broken into different modes that are employed either individually or in combination; the outcome of one mode of analysis may inform the application of other modes.

Data reduction This is the process of converting *raw* data into something more useful or meaningful to the experimenter: for example, converting the voltage changes in a particle detector (e.g. a proportional counter) into the records of the times and energies of individual particle detections. In turn, these may be further reduced into an energy spectrum for a specific type of particle.

[4] The earliest reference to this phrase I can find is Bailey (1967, ch. 2, p. 23).

[5] 'Data' is the plural of 'datum' and means 'items of information', although it has now become acceptable to use 'data' as a singular mass noun rather like 'information'.

Exploratory data analysis (EDA) is an approach to data analysis that uses quantitative and graphical methods in an attempt to reveal new and interesting patterns in the data. One does not test a particular hypothesis, but instead 'plays around with the data', searching for patterns suggestive of new hypotheses.

Inferential data analysis Sometimes known as 'confirmational data analysis'. We can divide this into two main tasks: model checking and parameter estimation. The former is the process of choosing which of a set of models provides the most convincing explanation of the data; the latter is the process of estimating values of a model's unknown parameters.

Exploratory data analysis is all about summarising the data in ways that might provide clues about their nature, and inferential data analysis is about making reasonable and justified inferences based on the data and some set of hypotheses.

1.5 Random samples

Our data about the real world are almost always incomplete, affected by random errors, or both. Let's say we wanted to find the answer to some important question: does the UK population prefer red or green sweets? We could survey the entire population and in principle get a complete answer, but this would normally be impractical. So we settle for a subset of the population, and assume this is representative of the population at large. Our results from the subset of people we actually survey is a *sample* and this is drawn from some *population* (of all the responses from the entire population). The sample is just one of the many possible samples that could be obtained from the same population.

But what we're interested in is the population, so we need to use what we know about the sample to infer something about the population. A small sample is easy to collect, but smaller samples are also more susceptible to random fluctuations (think of surveying just one person and extrapolating his/her answer to the entire population); a larger sample is less prone to such fluctuations but is also harder to collect. We also need to be sure to sample randomly and in an unbiased fashion – if we only sample younger people, or people in certain counties, these may not reflect the wider population. We need ways to quantify the properties of the sample, and also to quantify what we can learn about the population. This is statistics.

You may be left thinking: what's this got to do with experiments in the physical sciences? We often don't have a simple population from which we pull a random sample. Each time we perform some measurement (or series of measurements) we are collecting a sample of possible data. We can think of our sample as being drawn from a population, a hypothetical population of all the possible data that could be

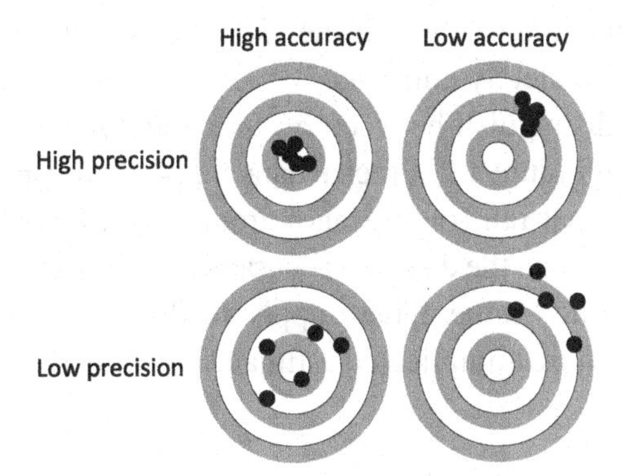

Figure 1.2 Illustration of the distinct concepts of accuracy and precision as applied to the positions of 'shot' on a target.

produced from our measurement(s). The differences between samples are due to randomness in the experiment or measurement processes.

1.5.1 Errors and uncertainty

The type of randomness described above is usually called *random error* (or measurement error) by physicists (the term *error* is used differently by statisticians[6]). Here, *error* does not mean a mistake as in the usual sense. To most scientists the 'measurement error' is an estimate of the repeatability of a measurement. If we take some data and use them to infer the speed of sound through air, what is the error on our measurement? If we repeat the entire experiment – under almost identical conditions – chances are the next measurements will be slightly different, by some unpredictable amount. As will further repeats. The 'random error' is a quantitative indication of how close repeated results will be. Data with small errors are said to have high *precision* – if we repeat the measurement the next value is likely to be very close to the previous value(s).

In addition to random errors, there is another type of error called *systematic error*. A systematic error is a bias in a measurement that leads to the values being systematically either too low or too high, and may arise from the selection of the sample under study or the calibration of the instrument used. Data with small systematic error are said to be *accurate*; if only we could reduce the random error we could get a result extremely close to the 'true' value. Figure 1.2 illustrates the difference between precision and accuracy. The experimenter usually works to reduce the impact of both random and systematic errors (by 'beating down the

[6] To a statistician, 'error' is a technical term for the discrepancy between what is observed and what is expected.

errors') in the design and execution of the experiment, but the reality is that such errors can never be completely eliminated.

It is important to distinguish between *accuracy* and *precision*. These two concepts are illustrated in Figure 1.2. Precise data are narrowly spread, whereas accurate data have values that fall (on average) around the true value. Precision is an indicator of variation within the data and accuracy is a measure of variation between the data and some 'true' value. These apply to direct measurements of simple quantities and also to more complicated estimates of derived quantities (Chapters 6 and 7).

1.6 Know your data

There are several types of data you may be confronted with. The main types are as follows.

Categorical data take on values that are not numerical but can be placed in distinct categories. For example, records of gender (male, female) and particle type (electron, pion, muon, proton etc.) are categorical data.

Ordinal data have values that can be ranked (put in order) or have a rating scale attached, but the differences between the ranks cannot be compared. An example is the Likert-type scale that you see on many surveys: 1, strongly disagree; 2, disagree; 3, neutral; 4, agree; 5, strongly agree. These have a definite order, but the difference between options 1 and 2 might not be the same as between options 3 and 4.

Discrete data have numerical values that are distinct and separate (e.g. 1, 2, 3, ...). Examples from physics might be the number of planets around stars, or the number of particles detected in a certain time interval.

Continuous data may take on any value within a finite or infinite interval. You can count, order and measure continuous data: for example, the energy of an accelerated particle, temperature of a star, ocean depth, magnetic field strength etc.

Furthermore, data may have many dimensions.

Univariate data concern only one variable (e.g. the temperature of each star in a sample).

Bivariate data concern two variables (e.g. the temperatures and luminosity of stars in a sample). Each data point contains two values, like the coordinates of a point on a plane.

Multivariate data concern several variables (e.g. temperature, luminosity, distance etc. of stars). Each data point is a point in an N-dimensional space, or an N-dimensional vector.

As mentioned previously, there are two main roles that variables play.

Explanatory variables (sometimes known as independent variables) are manipulated or chosen by the experimenter/observer in order to examine change in other variables.

Response variables (sometimes known as dependent variables) are observed in order to examine how they change as a function of the explanatory variables.

For example, if we recorded the voltage across a circuit element as we drive it with different AC frequencies, the frequency would be the explanatory variable, and the response variable would be the voltage. Usually the error in the explanatory variable is far smaller than, and can be neglected by comparison with, the error on the response variables.

1.7 Language

The technical language used by statisticians can be quite different from that commonly used by scientists, and this language barrier is one of the reasons that science students (and professional researchers!) have such a hard time with statistics books and papers. Even within disciplines there are disagreements over the meaning and uses of particular terms.

For example, physicists often say they *measure* or even *determine* the value of some physical quantity. A statistician might call this *estimation*. Physicists tend to use words like *error* and *uncertainty* interchangeably and rather imprecisely. In these cases, where conventional statistical language or notation offers a more precise definition, we shall use it. This is a deliberate choice. By using terminology and notation more like that of a formal statistics course, and less like that of an undergraduate laboratory manual, we hope to give the readers more scope for using and developing their knowledge and skills. It should be easier to understand more advanced texts on aspects of data analysis or statistics, and understand analyses from other fields (e.g. biology, medicine).

This means that we do not explicitly make use of the definitions set out in the *Guide to the Expression of Uncertainty in Measurement* (GUM, 2008). The document (now with revisions and several supplements) is intended to establish an industrial standard for the expression of uncertainty. Its recommendations included categorising uncertainty into 'type A' (estimated based on statistical treatment of a sample of data) and 'type B' (evaluated by other means), using 'standard uncertainty' for the standard deviation of an estimator, 'coverage factor' for a multiplier on the 'combined standard uncertainty'. And so on. These recommendations may be valuable within some fields such as metrology, but they are not standard in most physics laboratories (research or teaching) as of 2013, and are unlikely to be taken up by the broader community of researchers using and researching statistics and data analysis.

1.8 Statistical computing using R

You will need to be able to use a computer to do statistical data analysis on all but the smallest datasets. It is still possible to understand the ideas and methods of statistical data analysis in purely theoretical terms, without learning how to perform the analysis using a computer. The purpose of this book is to help you not only understand and interpret simple statistical analyses, but also perform analyses on data, and that means using a computer.

Throughout this book we give examples of statistical computing using the R environment (see Appendix A). R is an environment for statistical computation and data analysis. It is really a programming language with an integrated suite of software for manipulating data, producing plots and performing calculations, and has a very wide range of powerful statistical tools 'built in'. Using R it is relatively simple to perform statistical calculations accurately – this means you can spend less time worrying about the computational details, and more time thinking about the data and the statistical concepts. Appendix A provides a gentle introduction and a walkthrough of R.

Throughout the text are shaded boxes (R.boxes) containing the R code to carry out or demonstrate the procedures discussed in the accompanying text. Lines of R are written with `typewriter font`; these are meant to be typed at the R command line. As you progress through the book, working through the examples of R code, you will acquire the skills necessary to complete the data analysis case studies (and hopefully more besides). Of course, R is just one of the options you have for carrying out statistical computing. If your preferences lie elsewhere you should still be able to gain from the book by skipping past the R.boxes, or translating their contents into your favourite computing language.

1.9 How to use this book

This book is intended to provide a reasonably self-contained introduction to designing, performing and presenting statistical analyses of experimental data. Several devices are used to encourage you, the reader, to engage with the material rather than *just* read it. When a new term is used for the first time it usually appears in *italics* and is then defined, and to aid your memory there is a glossary of statistical terms towards the back of the book, along with a crib sheet for the mathematical notation. Dotted throughout the notes are two types of text box: white boxes contain examples or applications of ideas discussed in the text; shaded boxes ('R.boxes') contain examples using the R computing environment for you to work through yourself. We rely heavily on examples to illustrate the main ideas, and these are based on real data. The datasets are discussed in Appendix B.

In outline, the rest of the book is organised as follows.

- Chapter 2 discusses numerical and graphical summaries of data, and the basics of exploratory data analysis.
- Chapter 3 introduces some of the basic recipes of statistical analyses, such as looking for difference of the mean, or estimating the gradient of a straight line relationship.
- Chapter 4 introduces the concept of probability, starting with discrete, random events. We then discuss the rules of the probability calculus and develop the theory of random variables.
- Chapter 5 extends the discussion of probability to discuss some of the most frequently encountered distributions (and also mentions, in passing, the central limit theorem).
- Chapter 6 discusses the fitting of simple models to data and the estimation of model parameters.
- Chapter 7 considers the uncertainty on the parameter estimates, and model testing (i.e. comparing predictions of hypotheses to data).
- Chapter 8 discusses Monte Carlo methods, computer simulations of random experiments that can be used to solve difficult statistical problems.
- Appendix A describes how to get started in the computer environment R used in the examples throughout the text.
- Appendix B introduces the data case studies used throughout the text.
- Appendix C provides a refresher on combinations and permutations.
- Appendix D discusses the construction of confidence intervals (extending the discussion from Chapter 7).
- A glossary can be found on p. 217.
- A list of the notation can be found on p. 224.

2

Statistical summaries of data

The greatest value of a picture is when it forces us to notice what we never expected to see.

John Tukey (1977),
statistician and pioneer of exploratory data analysis

How should you summarise a dataset? This is what descriptive statistics and statistical graphics are for. A *statistic* is just a number computed from a data sample. Descriptive statistics provide a means for summarising the properties of a sample of data (many numbers or values) so that the most important results can be communicated effectively (using few numbers). Numerical and graphical methods, including descriptive statistics, are used in *exploratory data analysis* (EDA) to simplify the uninteresting and reveal the exceptional or unexpected in data.

2.1 Plotting data

One of the basic principles of good data analysis is: *always plot the data.* The brain–eye system is incredibly good at recognising patterns, identifying outliers and seeing the structure in data. Visualisation is an important part of data analysis, and when confronted with a new dataset the first step in the analysis should be to plot the data. There is a wide array of different types of statistical plot useful in data analysis, and it is important to use a plot type appropriate to the data type. Graphics are usually produced for screen or paper and so are inherently two dimensional, even if the data are not.

The variables can often be classified as *explanatory* or *response*. We are usually interested in understanding the behaviour of the response variable as a function of the explanatory variable, where the explanatory variable is usually controlled by the experimenter. Different plots are suitable depending on the number and type of the response variable.

- Data with one variable (univariate)
  - If the data are continuous, we can make a *histogram* showing how the data are distributed. A smooth *density curve* is an alternative to a histogram.
  - If the data are discrete or categorical, we could produce a *bar chart*, similar to a histogram but with gaps between the bars to indicate their discreteness.
  - If the data are a time series (a series of points taken at distinct times), we can make a *time series plot* by marking them as points on the x–y plane with y the data and x the time corresponding to each data point.
  - If the data are fractions of a whole, we may use compositional plots such as the *pie chart*; however, these are rarely used in scientific and statistical graphics (it is usually more efficient to present the proportions in a table or a bar chart).
- Data with two variables (bivariate)
  - If both variables are continuous, we may use a *scatter plot* where the data are plotted as points on the x–y plane.
  - There are many ways of augmenting a standard scatter plot, such as joining the points with lines (if the order is important or if it improves clarity), overlaying a smoothed curve or theoretical prediction curve and including error bars to indicate the precisions of the measurements.
  - If the explanatory variable is discrete (or binned continuous), we may choose from a *dotchart*, *boxplot*, *stripchart* or others.
- Data with many variables (multivariate)
  - A matrix of several scatter plots, each showing a different pair of variables, may be used to illustrate the dependence of each variable upon each of the others.
  - A *coplot* shows several scatter plots of the same two variables, where the data in each panel of the plot differ by the values of a third variable.
  - With three continuous variables we can make a projection of the three-dimensional equivalent of the scatter plot.
  - Another variation on the three-dimensional scatter plot is the *bubble plot*, which uses differently sized symbols to represent a third variable.
  - If we have one response variable and two explanatory variables, we can make an image using either greyscale, colours or contours to indicate the values of the response variable over the explanatory dimensions, or we can construct a projection of the surface, e.g. $z = f(x, y)$.

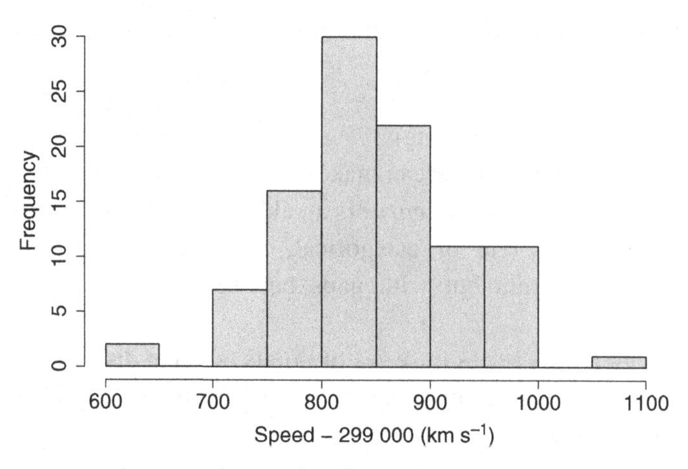

Figure 2.1 Histogram of the 100 Michelson speed-of-light data points.

2.2 Plotting univariate data

Michelson's data – see Appendix B, section B.1 – records 100 experimental values from his speed-of-light experiment. For compactness the tabulated data have had the leading three digits removed (i.e. 299 000 km s^{-1} subtracted). How should we plot these data? One option is an *index plot*, which plots points on the x–y plane at coordinates $(1, y_1)$, $(2, y_2)$ and so on, one point for each data value y_i. The order of the points is simply the order they occur in the table, which may (or may not) be the order they were obtained. Such a plot would make it much easier to see the 'centre' and 'spread' of the sample, compared with a table of raw numbers. But there are more revealing ways to view the data.

2.2.1 Histogram

One way to simplify univariate data is to produce a *histogram*. A histogram is a diagram that uses rectangles to represent frequency, where the areas of each rectangle are proportional to the frequencies. To produce a histogram one must first choose the locations of the *bins* into which the data are to be divided, then one simply counts the number of data points that fall within each bin. See Figure 2.1 (and R.box 2.1).

A histogram contains less information than the original data – we know how many data points fell within a particular bin (e.g. the 700–800 bin in Figure 2.1), but we have lost the information about which points and their exact values. What we have lost in information we hope to gain in clarity; looking at the histogram it is clear how the data are distributed, where the 'central' value is and how the data points are spread around it.

R.Box 2.1
Histograms

The R command to produce and plot a histogram is `hist()`. The following shows how to produce a basic histogram from Michelson's data (see Appendix B, section B.1):

```
hist(morley$Speed)
```

We can specify (roughly) how many histogram bins to use by using the `breaks` argument, and we can also alter the colour of the histogram and the labels as follows:

```
hist(morley$Speed, breaks=25, col="darkgray",
     main="", xlab="speed - 299,000 (km/s)")
```

This `hist()` command is quite flexible. See the help pages for more information (type `?hist`).

2.2.2 Bar chart

The *bar chart* is a relative of the histogram. Frequencies are indicated by the lengths of bars, which should be of equal width. Bar charts are used for discrete or categorical data, and a histogram is used for continuous data; neighbouring histogram bins touch each other, bar chart bars do not. For example, measurements of the speed of light are (in principle) continuous since the measured value can take any real number over some range, and so a histogram may be used. But if we were to plot data from a poll of support for different political parties, we should use a bar chart, since the data are categorical (different parties).

Figure 2.2 shows a bar chart for the data recorded by Rutherford and Geiger (see Appendix B, section B.2). The data record the number of intervals during which there were zero scintillations, one scintillation, two scintillations, up to 14 (there were no intervals with 15 or more scintillations). The data are discrete – the number of scintillations per interval, shown along the horizontal axis, must be an integer – and so a bar chart is appropriate.

R.Box 2.2
A simple bar chart

There are two simple ways to produce bar charts using R. Let's illustrate this using the Rutherford and Geiger data (see Appendix B, section B.2):

```
plot(rate, freq, type="h")
plot(rate, freq, type="h", bty="n",
```

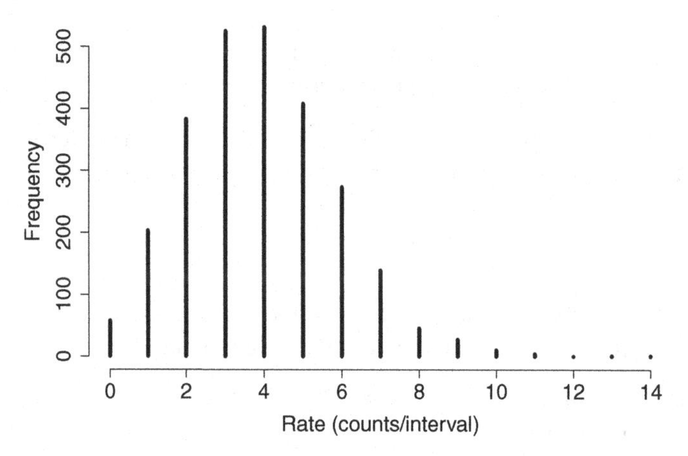

Figure 2.2 Bar chart showing the Rutherford and Geiger (1910) data of the frequency of alpha particle decays. The data comprise recordings of scintillations in 7.5 s intervals, over 2608 intervals, and this plot shows the frequency distribution of scintillations per interval.

```
       xlab="Rate (counts/interval)",
       ylab="Frequency", lwd=5)
```

The first line produces a very basic plot using the `type="h"` argument. The second line produces an improved plot with user-defined axis labels, thicker lines/bars and no box enclosing the data area. An alternative is to use the specialised command `barplot()`.

```
barplot(freq, names.arg=rate, space=0.5,
      xlab="Rate (cts/interval)",
      ylab="Frequency")
```

Here the argument `space=0.5` determines the sizes of the gaps between the bars, and `names.arg` defines the labels for the x-axis. If the data were categorical, we could produce a bar chart by setting the `names.arg` argument to the list of categories.

2.3 Centre of data: sample mean, median and mode

Probably the first conclusion we might draw from looking at Michelson's data is that the measured values lie close to 299 800 km s^{-1}. What we have just done is make a *numerical summary* of the data – if we needed to communicate the most important aspects of this dataset to a colleague in the smallest amount of information, a sensible place to start would be with a summary like this, which gives some idea of the 'centre' of the data. But instead of making a quick informal

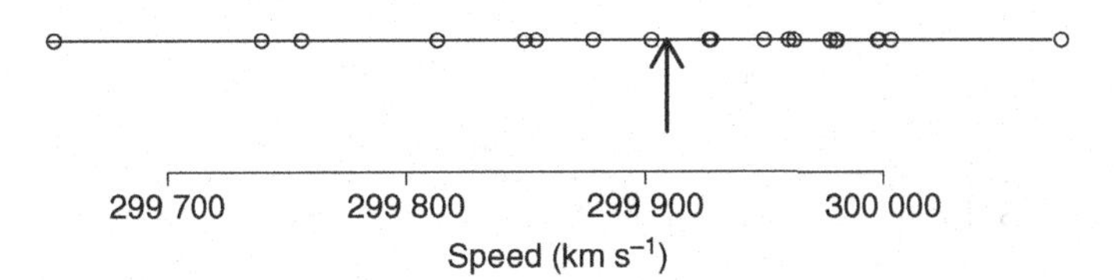

Figure 2.3 Illustration of the mean as the balance point of a set of weights. The data are the first 20 of the Michelson data points.

guess of the centre we could instead calculate and quote the *mean* of the sample, defined by

$$\overline{x} = \frac{1}{n}\sum_{i=1}^{n} x_i \tag{2.1}$$

where x_i ($i = 1, 2, \ldots, n$) are the individual data points in the sample and n is the size of the sample. If x are our data, then $\bar{x}$ is the conventional symbol for the sample mean. The sample mean is just the sum of all the data points, divided by the number of data points. Strictly, this is the *arithmetic* mean. The mean of the first 20 Michelson data values is 909 km s^{-1}:

$$\bar{x} = \frac{1}{20}(850 + 740 + 900 + 1070 + 930 + 850 + \ldots + 960) = 909.$$

One way to view the mean is as the balancing point of the data stretched out along a line. If we have n equal weights and place them along a line at locations corresponding to each data point, the mean is the one location along the line where the weights balance, as illustrated in Figure 2.3.

The mean is not the only way to characterise the centre of a sample. The sample *median* is the middle point of the data. If the size of the sample, n, is odd, the median is the middle value, i.e. the $(n + 1)/2$th largest value. If n is even, the median is the mean of the middle two values (the $n/2$th and $n/2 + 1$th ordered values). The median has the sometimes desirable property that it is not so easily swayed by a few extreme points. A single outlying point in a dataset could have a dramatic effect on the sample mean, but for moderately large n one outlier will have little effect on the median. The median of the first 20 light speed measurements is 940 km s^{-1}, which is not so different from the mean – take a look at Figure 2.1 and notice that the histogram is quite symmetrical about the mean.

The last measure of the centre we shall discuss is the sample *mode*, which is simply the value that occurs most frequently. If the variable is continuous, with no repeating values, the peak of a histogram is taken to be the mode. Often there is more than one mode; in the case of the 100 speed of light values, there are two values that occur most frequently (810 and 880 km s^{-1} occur 10 times each). Once

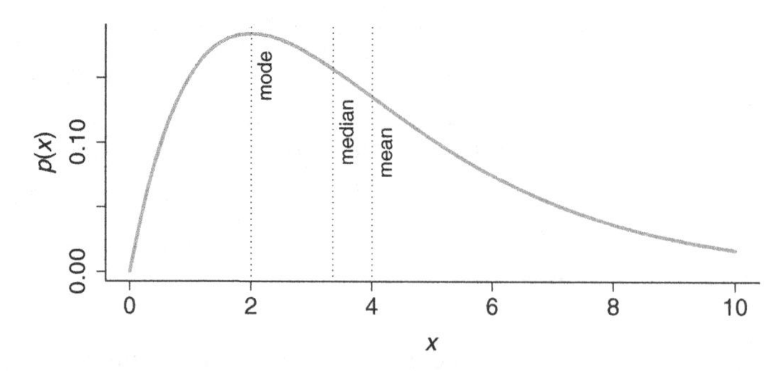

Figure 2.4 Illustration of the locations of the mean, median and mode for an asymmetric distribution, $p(x)$, where x is some random variable.

we bin the Michelson data into a histogram it becomes clear that the distribution has a single mode around 800–850 km s^{-1} (see Figure 2.1).

Now we have three measures of centrality, but the one that is used the most is the mean, often just called the average. If we have some theoretical distribution of data spread over some range, we may calculate the mean, median and mode using methods discussed in Chapter 5.

Figure 2.4 illustrates how the three different measures differ for some theoretical distribution. The mean is like the centre of gravity of the distribution (if we imagine it to be a distribution of mass density along a line); the median is simply the 50% point, i.e. the point that divides the curve into halves with equal areas (equal mass) on each side; the mode is the peak of the distribution (the densest point). If the distribution is symmetrical about some point, the mean and median will be the same, and if it is symmetrical about a single peak then the mode will also be the same, but in general the three measures differ.

R.Box 2.3

Mean, median and mode in R

We can use R to calculate means and medians quite easily using the appropriately named `mean()` and `median()` commands. The variable `morley$Speed` contains the 100 speed values of Michelson. To calculate the mean and median, and add on the offset (299 000 km s^{-1}), type

```
mean(morley$Speed) + 299000
median(morley$Speed) + 299000
```

The modal value is not quite as easy to calculate as the mean or median since there is no built-in function for this. One simple way to find the mode is to view a histogram of the data and select the value corresponding to the peak.

Box 2.1
Different averages

Imagine a room containing 100 working adults randomly selected from the population. Then Bill Gates walks into the room. What happens to the mean wealth of the people in the room? What about the median or mode? These different measures of 'centre' react very differently to an extreme outlier (such as Bill Gates). What will happen to the average height (mean, median and mode) of the people in the room if the world's tallest man walks in?

What is the average number of legs for an adult human? The mode and the median are surely two, but the *mean* number of legs is slightly less than two!

2.4 Dispersion in data: variance and standard deviation

The sample mean is a very simple and useful single-number summary of a sample, and it gives us an idea of the typical location of the data. If we required slightly more information about the sample a good place to start would be with some measure of the spread of the data around this central location: the dispersion around the mean. We could start by calculating the mean of the deviations between each data value and the sample mean. But this is useless as it always equals zero. Take another look at the definition for the sample mean (equation 2.1) and notice how the sample mean is the one value that ensures the (data – mean) deviations sum to zero (recall the balance of Figure 2.3):

$$\frac{1}{n}\sum_{i=1}^{n}(x_i - \bar{x}) = \frac{1}{n}\sum_{i=1}^{n}x_i - \frac{1}{n}\sum_{i=1}^{n}\bar{x} = \bar{x} - \frac{n}{n}\bar{x} = \bar{x} - \bar{x} = 0. \tag{2.2}$$

The negative deviations exactly cancel the positive deviations.

If instead we square the deviations, then all the elements of the sum are positive (or zero), so the average of the squared deviation seems like a more useful measure of the spread in a sample. The sample variance is defined as

$$s_x^2 = \frac{1}{n-1}\sum_{i=1}^{n}(x_i - \bar{x})^2. \tag{2.3}$$

This is almost the mean of the squared deviations. But notice that we have divided by $n - 1$ rather than n: the story behind this is sketched out in the box. Table 2.1 illustrates explicitly the steps involved in calculating the variance using the first 20 values from the Michelson dataset: first we compute the sample mean, then subtract this from the data, and compute the sum of the squared $data - mean$ deviations. Of course, in real data analysis this calculation is always performed by computer.

Table 2.1 *Illustration of the computation of variance using the first $n = 20$ data values from Michelson's speed of light data. Here x_i are the data values, and the sample mean is their sum divided by n: $\bar{x} = 18\,180/20 = 909\,\text{km s}^{-1}$. The $x_i - \bar{x}$ are the deviations, which always sum to zero. The squared deviations are positive (or zero) valued and sum to a non-negative number. The sum of squared deviations divided by $n - 1$ gives the sample variance:* $s^2 = 209\,180/19 = 11\,009.47\,\text{km}^2\,\text{s}^{-2}$.

i	1	2	3	4	5	...	20	sum
Data x_i (km s^{-1})	850	740	900	1 070	930	...	960	18 180
$x_i - \bar{x}$ (km s^{-1})	−59	−169	−9	161	21	...	51	0
$(x_i - \bar{x})^2$ (km^2 s^{-2})	3481	28 561	81	25 921	441	...	2601	209 180

The sample variance is always non-negative (i.e. either zero or positive), and will not have the same units as the data. If the x_i are in units of kg, the sample mean will have the same units (kg) but the sample variance will be in units of kg^2. The *standard deviation* is the positive square root of the sample variance, i.e. $s = \sqrt{s^2}$, and has the same units as the data x_i. Standard deviation is a measure of the typical deviation of the data points from the sample mean. Sometimes this is called the RMS: the root mean square (of the data after subtracting the mean).

Box 2.2

Why $1/(n-1)$ in the sample variance?

The sample variance is normalised by a factor $1/(n-1)$, where a factor $1/n$ might seem more natural if we want the mean of the squared deviations. As discussed above, the sum of the deviations $(x - \bar{x})$ is always zero. If we have the sample mean the last deviation can be found once we know the other $n - 1$ deviations, and so when we average the square deviation we divide by the number of independent elements, i.e. $n - 1$. This known as *Bessel's correction.*

Using $1/(n-1)$ makes the resulting estimate *unbiased*. Bias is the difference between an average statistic and the true value that it is supposed to estimate, and an unbiased statistic gives the right result when given a sufficient amount of data (i.e. in the limit of large n). For more details of the bias in the variance, see section 5.2.2 of Barlow (1989), or any good book on mathematical statistics.

The variance, or standard deviation, gives us a measure of the spread of the data in the sample. If we had two samples, one with $s^2 = 1.0$ and one with $s^2 = 1.7$, we would know the that the typical deviation (from the mean) is 30% times larger in the second sample (recall that $s = \sqrt{s^2}$).

R.Box 2.4
Variance and standard deviation

R has functions to calculate variances and standard deviations. For example, in order to calculate the mean, variance and standard deviation of the numbers $1, 2, \ldots, 50$:

```
x <- 1:50
mean(x)
var(x)
sd(x)
```

Likewise to calculate the variance of the entire Michelson sample

```
Speed <- morley$Speed
var(Speed)
```

The first line defines a new array in order to save us having to use the prefix `morley$...` every time we wish to access these data.

R.Box 2.5
Calculating with subarrays

If we want to calculate the variance for each of Michelson's five 'experiments' (each one is a block of 20 consecutive values) individually, we could use

```
mask <- morley$Expt == 2
mask
Speed[mask]
var(Speed[mask])
```

Note the use of the double equals sign (==) in testing for equality. The first line forms an array `mask`, the same size as the `Speed` array, with values that are `TRUE` where the condition is met (i.e. `Expt == 2`), and `FALSE` elsewhere. The third line forms a subarray from `Speed` by taking only those elements that occur where `mask` is `TRUE`). The third line shows how to compute the variance of this subset of the original data. We can repeat this process using a loop as follows:

```
for (i in 1:5) {
  print(var(Speed[morley$Expt==i]))
}
```

This looks quite complicated, so let's unpack it. The first part `for (i in 1:5)` {...} defines a loop. The second part (inside the curly brackets) defines what is to happens each time around the loop. The loop runs once for each of $i = 1, 2, 3, 4, 5$, and each time round it prints the variance of the sample of data with the corresponding experiment number i. The following may help illustrate the way loops are written in R:

```
for (i in 1:10) { print(i) }
```

2.5 Min, max, quantiles and the five-number summary

A simple two-point indicator of the spread of a data sample is the pair (minimum, maximum). Other measures of a sample commonly used in descriptive statistics are *quantiles*. The α quantile is simply the data point below which a fraction α of the data occur. The 0.25 quantile is then simply the value for which 25% of the data points are lower. The 0.5 quantile is the median. Some quantiles have special names, for example the 0.25, 0.5 and 0.75 quantiles are called the first, second and third quartiles, respectively. The median is the second quartile. The difference between the 0.75 and 0.25 quantiles is called the interquartile range (IQR). (Note that the first and third quartiles can be obtained by splitting the data about the median, and then finding the medians of the lower and upper halves.)

John Tukey (see Tukey, 1977) suggested a simple and compact five-number summary of a univariate dataset, now known as the Tukey five-number summary. This comprises the minimum, first quartile, median (second quartile), third quartile and maximum values of a sample. From these five numbers, one can get a reasonable impression of the way the data are distributed: the centre of the sample (median), the way the central 50% of the data are spread around the median (IQR) and the most extreme (lowest, highest) values in the sample.

R.Box 2.6

Tukey's five-number summary

There are two functions in R to calculate variations on Tukey's five-number summary. The first is

```
fivenum(0:100)
fivenum(Speed)
```

Here the reported values for the first, second (median) and third quartiles are given as the closest actual data values. There is a variation on this:

```
summary(0:100)
summary(Speed)
```

The two methods differ slightly in how the quartiles are calculated. Note that the `summary()` command calculates the mean for free.

2.6 Error bars, standard errors and precision

From the above, we now have some numerical and graphical ways to summarise data, and in particular its centre and spread. However, we still have not made any attempt to quantify how precise these summaries might be. There are 100 values in the Michelson datasets, divided into five experiments, each of 20 measurements. For each of the experiments, we can calculate a mean and variance for the 20 measurements. From these, we may calculate the *standard error* on the sample mean. Here it is:

$$\mathrm{SE}_{\bar{x}} = \sqrt{\frac{s_x^2}{n}} \tag{2.4}$$

which is just the square root of the sample variance, s_x^2 (equation 2.3), divided by the size of the sample, n. We shall not be concerned with where this formula comes from until later chapters. For now, we consider it a useful, simple, approximate formula for the uncertainty on the sample mean, $\bar{x}$.

What is the meaning of the standard error? Imagine repeating an experiment n times and, to get the 'best' result, taking the sample mean of the measurements, $\bar{x}_1$. We could repeat the whole set of n experiments and calculate another sample mean, $\bar{x}_2$, and so on. If we do this many times, we have a sample of mean values, $\bar{x}_j$, each of which is an independent estimate of the population ('true') mean, μ. The standard error is an estimate of the standard deviation of the sample means from the expected (population) mean value. In other words, we expect the sample means to be about one standard error from the population mean. Thus the standard error gives us an idea of the precision of the sample mean. You can see that as n increases, the standard error decreases; one would expect the precision of the mean to improve as more data are acquired. In statistics the word *precision* (see section 1.6) is sometimes used for the reciprocal of the variance of the data. The precision of the mean $\bar{x}$ is $1/\mathrm{SE}_{\bar{x}}^2$.

Let's look at the sample means and standard errors for the Michelson data divided into five 'experiments'. Figure 2.5 shows the sample means and their standard errors. The standard errors are illustrated by *error bars*, which run from $\bar{x} - \mathrm{SE}_{\bar{x}}$ to $\bar{x} + \mathrm{SE}_{\bar{x}}$. This figure summarises each of the five experiments in terms of two numbers each, the mean and a measure of its precision, and the five experiments can easily be compared with each other and the modern, accepted value.

R.Box 2.7
Standard errors in R

There is no single command to compute the standard error in R, but one may make use of the `var()` function to make the calculation simple. For example, to compute the mean, variance and standard error of the Michelson data

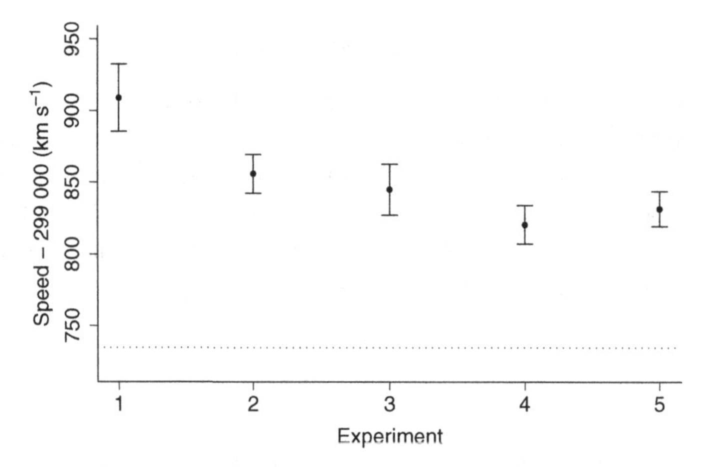

Figure 2.5 The sample means for each of the five 'experiments' of Michelson, each comprising 20 measurements. The standard errors for each mean are indicated by the error bars. Notice the sidebars at the end of each error bar. These help define the ends of each error bar, but may clutter the graphic when there are a lot of data to present. The dotted line shows the modern value for the speed of light in air. From this graphic, one can start to make inferences about Michelson's measurements.

```
x <- morley$Speed
mean(x)
var(x)
sqrt( var(x) / length(x) )
```

where the `length(x)` function returns the number of data points.

R.Box 2.8

Standard errors by group, part 1

It is possible to calculate a statistic (e.g. mean or variance) for each of the five experiments in an efficient manner by first re-organising the data into a matrix. Once this is done we can make use of some powerful matrix tools in R. In the following example, the speed data are converted to a matrix with 20 rows (and therefore five columns, one for each 'experiment') called `speed`.

```
speed <- matrix(morley$Speed, nrow=20)
speed
     [,1] [,2] [,3] [,4] [,5]
[1,]  850  960  880  890  890
[2,]  740  940  880  810  840
[3,]  900  960  880  810  780
```

```
[4,] 1070  940  860  820  810
[5,]  930  880  720  800  760
[6,]  850  800  720  770  810
...   ...  ...  ...  ...  ...
```

It is important to check that the matrix is arranged in the right way. Here we see all the data from first experiment in the first column – compare with the output of

```
morley$Speed[morley$Expt == 1]
```

R.Box 2.9

Standard errors by group, part 2

With the Michelson data arranged in a matrix, we can use the `apply()` command to apply any function, e.g. `mean()` or `var()`, to every row or column of the matrix. For example, to calculate the mean and variance of the data in each column, and then store the results in new data objects, we can use

```
speed.mean <- apply(speed, 2, mean)
speed.var <- apply(speed, 2, var)
speed.var
```

The command `apply(speed, 2, var)` takes the matrix called `speed` and applies the function `var()` to each of its columns to calculate the variance. You could also use `mean`, `sd`, `sum`, or any other valid R command. The second argument (i.e. `2`) specifies columns should be analysed. If instead we used `1`, we would get the variance over each row. This approach, applying the same function over rows or columns of an array, is usually faster (on large datasets) and more elegant than using loops.

R.Box 2.10

Standard errors by group, part 3

Finally, the standard errors for the five 'experiments' are just the square roots of these variances divided by the number of data points in each experiment. We find the number of data points in each column using the command `apply()` to apply the `length()` function (we know the answer is 20).

```
speed.n <- apply(speed, 2, length)
se <- sqrt(speed.var / speed.n)
se
data.frame(speed.mean, speed.var, speed.n, se)
```

Remember that R is case sensitive, so `se` is not the same object as `SE`. The last line uses the four new vectors (of the means, variance, lengths and standard errors) as columns of a new object, a *data frame* (similar to a matrix but the columns may be formed from different types of data).

R.Box 2.11
Plotting error bars

There are several ways to add error bars to a graphic in R. One way is using the `segments()` command to draw a series of line segments between $x-$ *error* and $x+$ *error*. If we have sample means with standard errors (as in the previous box), we may plot them as follows:

```
Expt <- 1:length(speed.mean)
plot(Expt, speed.mean, ylim=c(780,950), pch=16,
     bty="l", xlab="Experiment",
     ylab="Speed - 299,000 (km/s)")
segments(Expt, speed.mean-se, Expt, speed.mean+se)
```

where the second line plots the data and the third line adds the error bars. The `segments` command takes as its input `segments(x0,y0,x1,y1)` and draws lines between coordinates (`x0,y0`) and (`x1,y1`). A variation on this is to use the `arrows` command to give each error bar a sidebar (as in Figure 2.5):

```
arrows(Expt, speed.mean-se, Expt, speed.mean+se,
       code=3, angle=90, length=0.1)
```

Where the first four arguments give the coordinates of the endpoints (as for the `segments()` command), and the last three define two-sided arrows (`code=3` means draw an arrow head at both ends of the arrow), with flat arrow heads (`angle=90`) and the extent of the arrow heads (`length=0.1`).

It is common in physical science to expect error bars accompanying data whenever appropriate; they immediately allow the viewer to gauge the precision of the estimate or measurement. What use is an estimate without any measure of how reliable it is?

2.7 Plots of bivariate data

2.7.1 Scatter plot

So far we have considered only data that are records of the values of a single variable, such as Michelson's speed of light measurements. However, a great deal of data analysis concerns data with more than one variable, often one or more *response* variable, observed or measured at different values of one or more *explanatory* variables.

R.Box 2.12
Scatter plots in R

The R command `plot()` will produce a basic scatter plot from two (equal length) arrays of numbers. The *Hipparcos* data shown in Figure 2.6 are described in Appendix B (section B.4). Using the reduced data file `hip_clean.txt` we can produce a simple plot

```
hip <- read.table("hip_clean.txt", header=TRUE)
```

This creates a data array called `hip` that contains the contents of the file: 14 columns and 5740 rows of data. A simple scatter plot may be produced using

```
plot(hip$BV, hip$V)
```

However, with a little more effort we can do much better than this.

The simplest way to visualise data with two continuous variables is a *scatter plot*, where each data point (pair of numbers) is treated as a coordinate and is marked with a symbol on the x–y plane. Scatter diagrams are used to reveal relationships between pairs of variables, and are among the most widely used diagrams in all of science. They can be enormously powerful; indeed, some of the most important diagrams and relations in science were discovered by examination of scatter plots.

Figure 2.6 shows one such example from astronomy. This is a *Hertzsprung–Russell* diagram (sometimes known as a colour–magnitude diagram) and shows the luminosity against colour index for a sample of nearby stars. Each point represents a star, the horizontal position of the points represents the $B - V$ colour index (a simple measure of the colour of the star, which depends on its temperature), and the vertical position represents the absolute magnitude (an upside-down and logarithmic measure of the luminosity). When these two variables are used to construct a scatter diagram for a sample of stars, it is clear there is a great deal of structure in the data, patterns that would not be at all obvious by examination of a table of numbers, or of graphical examination of either variable separately.

R.Box 2.13
Basic scatter plot design

The following command shows how to produce a better scatter plot:

```
plot(hip$BV, hip$V.abs, pch=1, cex=0.5, bty="n",
    ylim=c(16, -3), xlim=c(-0.3, 2.0),
    ylab="V.abs (mag)", xlab="B-V (mag)")
```

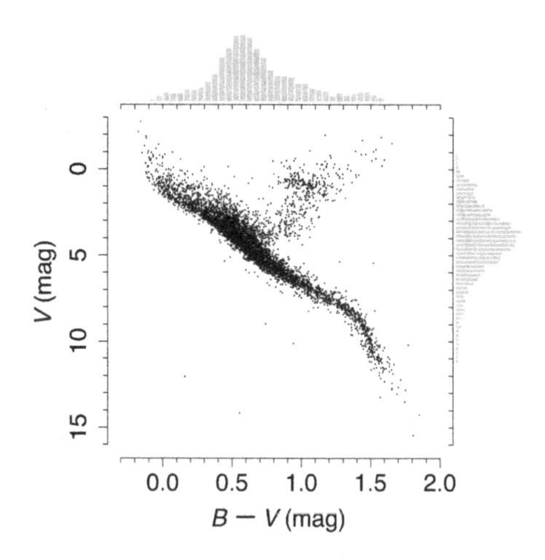

Figure 2.6 Example of a scatter plot showing data on 5740 stars using data from the *Hipparcos* astronomy satellite. Plotted is the V-band (green) absolute (distance corrected) magnitude against the $B - V$ colour index (difference between B and V-band magnitudes, a blue–green colour). Each point represents a star: brighter (smaller magnitude) stars are at the top, bluer stars are on the left. The plot clearly reveals structure in the data: most stars fall in the band from top left to bottom right, with a small island in the top right. This type of diagram is of fundamental importance in stellar astrophysics. For comparison we also show the histograms of each of the two variables (V and $B - V$) separately. The structure in the data is only apparent when looking at the two variables together using a scatter plot.

Here we have plotted V_{abs}, the absolute magnitude stored in the `V.abs` column (not the apparent magnitude in the `V` column), against $B - V$. The `pch=1` argument selects a plot symbol (1 is a hollow circle); `cex=0.5` makes the symbols smaller than default. A small, hollow symbol was chosen here to reduce the clutter from the large number of points to be plotted.

The option `ylim=c(16, -3)` sets the range of the vertical axis to run from 16 at the bottom to −3 at the top. The `xlim` argument is used to control the horizontal axis span. The arguments `xlab` and `ylab` are for setting the axis labels, and finally `bty="n"` defines what type of box to enclose the plot in (`"n"` means no box).

For more information on the arguments that can be changed within the `plot()` command, try `?plot` and `?par`.

How does one decide which observable to plot on the horizontal axis, and which on the vertical axis? In an experiment one usually studies the response of some variable(s) to changes in experimenter-controlled explanatory variables, in which case the explanatory variable is plotted along the horizontal axis and the response variable plotted along the vertical axis. However, it is often the case that neither variable is obviously an explanatory variable. For example, if we recorded

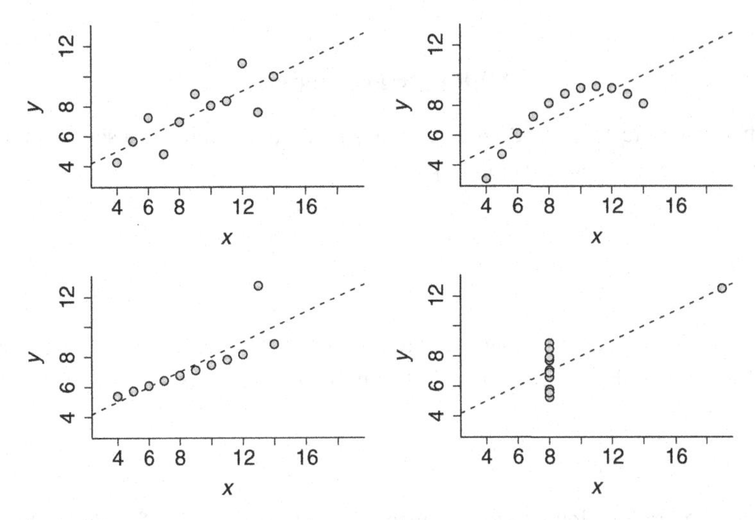

Figure 2.7 Scatter plots of four datasets. The data are shown with circles and the models shown with a dotted line. All four datasets are described by exactly the same means and variances (for each x and y), and simple analysis yields the same linear model for each dataset. Yet simply plotting the data like this reveals very different patterns in each. These are often known as 'Anscombe's quartet' (from a classic paper by Anscombe, 1973), and have been used by many authors (e.g. Tufte, 1986) to demonstrate the power of simple graphics to reveal structure in the data.

the orbital speeds and masses of comets, neither of these can be chosen by the experimenter/observer; neither are explanatory. In this case, it is a largely matter of preference which way round to draw the diagram.

Figure 2.7 demonstrates the power of the scatter diagram to reveal the structure in the data. This shows four datasets, each one is a sample of $n = 11$ points. For each dataset, the values of the variables all have the same mean, variance, etc., and they even give the same 'straight line' fit (Chapter 3). From the numerical summaries alone it is difficult to tell them apart. But differences are easy to see in scatter diagrams. This illustrates an important message for data analysis, one that is worth repeating: *always plot the data.*

2.7.2 Scatter plot extras

It is also possible to augment the scatter plot with additional graphics. For example, a *rug* is a series of tickmarks along one axis at the positions of each data point (R.box 2.14). Alternatively, we could show the histograms for each variable along the appropriate axes as in Figure 2.6. These distributions, where one or more variables are ignored and the dimensionality of the data is reduced, are called *marginal distributions*.

R.Box 2.14
Adding 'rugs' to plots

Let us first generate some random x, y values and plot them using a scatter plot. (We shall discuss random data generation later.)

```
x <- rchisq(50, df=2)
y <- x + 0.7*rnorm(50)
plot(x, y, pch=16)
```

Then we may use the rug() command to plot a rug along each axis – a rug simply plots a tickmark at the location of each data point along one axis.

```
rug(x, side=1)
rug(y, side=2)
```

This type of diagram allows us to visualise the joint distribution of x and y along with the individual (marginal) distributions of x and y considered separately. We can also add a rug to a histogram using the rug command (try it using the Michelson data).

Another possibility is to overlay a smoothed curve. There are a range of methods for generating smoothed curves from scattered data points: the simplest is to take the 'running mean' of every few data points; the mean will show less scatter than the raw data. More sophisticated methods may be more robust to outliers, or better suited to data of a particular type. Such overlays can be useful for drawing the eye to the way the response variable typically changes with the explanatory variable.

R.Box 2.15
Adding smoothed curves to scatter plots

We can overlay a curve on a scatter plot using the lines() function. The lowess() function is one of many functions (some quite sophisticated) for smoothing data, that is, it will produce a smoothed curve that runs through the centre of the data. The output of the lowess() command is a list containing the x and y values of the smoothed data, ready to be plotted. For example, using the random x, y data from above:

```
smooth.data <- lowess(x, y, f=0.4)
lines(smooth.data, lwd=4, col="blue")
```

where the lwd and col arguments to the lines() function set the width and colour of the line. The f argument to the lowess() function defines the degree of smoothing; a larger number gives a smoother result. Try adding a smoothed curve to the HR diagram (R.box 2.12).

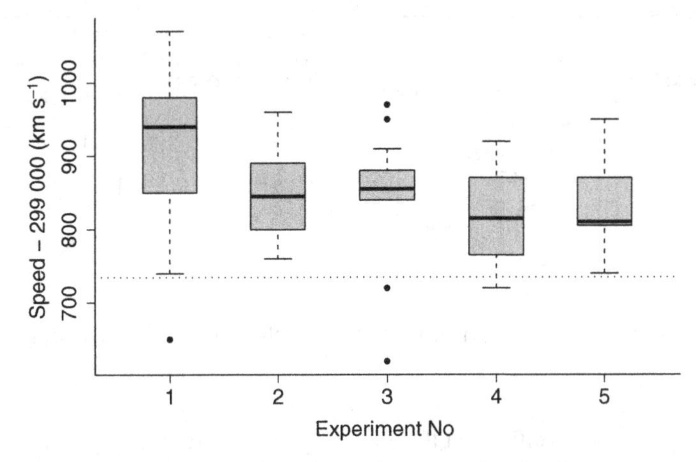

Figure 2.8 Example of a boxplot illustrating data for the five separate 'experiments' (each comprising 20 'runs') of the Michelson data taken in 1879 (Michelson, 1882; Stigler, 1977). The dotted line is the modern value for the speed of light in air. Compare with Figure 2.1.

2.7.3 Box plots and dot charts

A *boxplot* is a visual summary based on Tukey's five-number summary. It shows the first, second (median) and third quartiles by a box, with *whiskers* extending out no further than $1.5\times$IQR in each direction (to the most distant data point within this range). Any data points outside this range – the outliers – are individually plotted.

A boxplot can be used to quickly compare two or more distributions; it effectively allows several histograms to be compared on a common scale. It is useful for data with a continuous response and a discrete or categorical explanatory variable. Figure 2.8 compares the data from the five separate runs of Michelson's speed-of-light experiment, where each experiment comprises 20 individual measurements. In this case the experiment number is the explanatory variable and the speed measurement is the response variable.

R.Box 2.16
Boxplot

The R command to generate a boxplot is, not too surprisingly, `boxplot()`. The following example, again using the Michelson data, illustrates its use:

```
plot(morley$Exp, morley$Speed,
     bty="n", xlim=c(0.5, 5.5))
boxplot(morley$Speed ~ morley$Expt, add=TRUE)
```

The first line produces a simple scatter plot. The second overlays (using the `add=TRUE` argument) the boxplot. The first argument is a *formula* (`morley$Speed ~`

Figure 2.9 The power of statistical graphics. (Credit: xkcd.com)

morley$Expt) that specifies that the data to be plotted are morley$Speed and they should be divided into boxes according to morley$Expt. We can do better by specifying the axis labels, using shaded boxes (col), and thicker lines for the medians (medlwd):

```
boxplot(morley$Speed ~ morley$Expt,
        xlab="Experiment No.",
        ylab="speed - 299,000 (km/s)",
        medlwd=4, col="light grey")
```

The dot chart was designed by Cleveland (1985) as an alternative to the bar chart for plotting values of a continuous variable against categorical variable. An example is shown in Figure 2.10. This shows estimates of the masses of pulsars. The pulsar mass is a continuous variable, and there is one estimate for each pulsar, with the pulsar name a categorical variable. The faint horizontal lines reduce the confusion when connecting each data point with its label on the left – this becomes more important as more data are plotted. The dot chart is slightly more efficient in its use of ink (amount of black ink per data point) than an equivalent bar chart.

R.Box 2.17

Dot chart

The command to produce a dot chart in R is dotchart(). One simply enters a numerical vector as the first argument, and optionally a list of labels. For example, to plot the sizes of the planets first define the data:

```
d <- c(0.382, 0.949, 1.00, 0.532,
       11.209, 9.449, 4.007, 3.883)
```

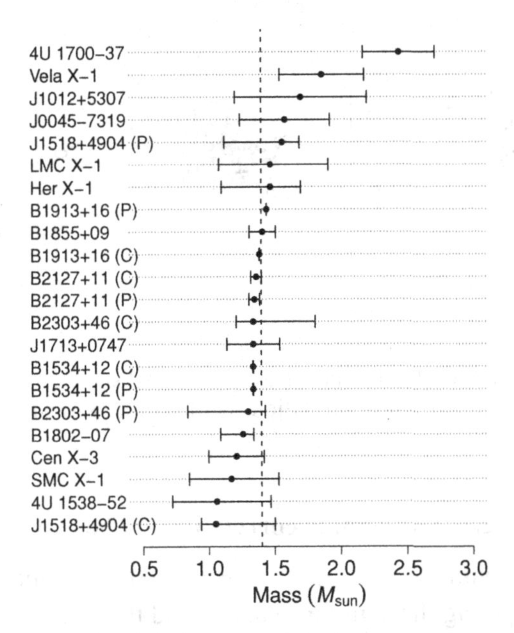

Figure 2.10 Example of a dot plot created with the `dotchart()` command in R. The data show the mass estimates for a sample of neutron stars ('P' and 'C' indicate the primary and secondary members of binary systems, respectively). Each line represents one neutron star, and each mass has an accompanying error bar. The vertical dotted line marks the theoretical lower limit for neutron star mass (1.4 $M_{\odot}$). The shape of the data – 22 lines, each with a text label – works well in portrait orientation. Data from Charles and Coe (2006).

```
names <- c("Mercury", "Venus", "Earth", "Mars",
           "Jupiter", "Saturn", "Uranus", "Neptune")
dotchart(d, names)
```

If we want to change the order so that the sizes of the planets increase from bottom to top, we may use the `order()` function to change the orders (remembering to do the same to the size and label vectors!)

```
xlab <- expression(Diameter ~ (R[Earth]))
indx <- order(d)
dotchart(d[indx], names[indx],
         xlab=xlab, pch=16, cex=2)
```

We have also used `cex` and `pch` to change the character size and type. The `expression()` function is used to format the text of the label for the horizontal axis (it can be used to convert mathematical expressions into text for plotting). For more on writing mathematical expressions type `?plotmath`.

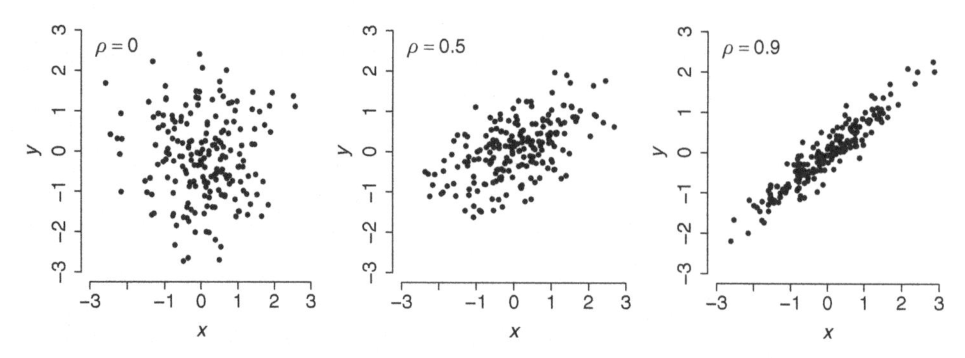

Figure 2.11 Examples of correlated data. Each panel shows $n = 200$ data points generated using different (true) correlation coefficients ρ. From left to right: $\rho = 0$, 0.5, 0.9.

2.8 The sample correlation coefficient

Given some data x_i (with $i = 1, 2, \ldots, n$) we can learn a lot about how the data are distributed by computing the sample mean, $\bar{x}$, and the sample variance, s_x^2. If we have bivariate data, for example data pairs x_i, y_i we can define the sample mean and variance of x_i, and of y_i, separately. But these do not tell us anything about how the y_i values change with the x_i values, the *joint* distribution of the x_i, y_i pairs.

There is a simple and useful way to quantify the way y changes with x. We can consider a sum of terms such as $(x_i - \bar{x})(y_i - \bar{y})$. This is the basis of the sample covariance s_{xy}:

$$s_{xy} = \frac{1}{n-1} \sum_{i=1}^{n} (x_i - \bar{x})(y_i - \bar{y}) . \tag{2.5}$$

This has the same units as $x \times y$. Compare this with the sample variance of equation 2.3, and it should be obvious that $s_{xx} = s_x^2$. The sample covariance is the sum over all data points of the product of deviations in x (from its mean) and deviations in y (from its mean). If y tends to increase when x increases, there will be more positive than negative terms in the sum, and the result will tend to be positive. If y tends to decrease with increases in x, the result will tend to be negative. If, on the other hand, deviations in y are just as likely to be positive or negative as x increases, then the sum will be close to zero. See Figure 2.11 for an illustration.

If we take the sample covariance and normalise it using the variances of the x and y, we form the correlation coefficient r:

$$r = \frac{s_{xy}}{s_x s_y} = \frac{1}{n-1} \sum_{i=1}^{n} \left(\frac{x_i - \bar{x}}{s_x} \right) \left(\frac{y_i - \bar{y}}{s_y} \right) . \tag{2.6}$$

Figure 2.12 Correlation doesn't mean causation. (Credit: xkcd.com)

This is a dimensionless statistic. There are many ways to rearrange the terms in this equation, but the above version shows how r is a product of the normalised deviations in x and in y. If x and y are so tightly connected that they only differ by an offset and a scaling factor, that is $y \propto x + c$, then $r = 1$ (the proof of this is left as an exercise). In fact, r is restricted to the range $[-1, 1]$: values close to 0 mean that y is as likely to increase as to decrease with increasing x, values close to 1 (or -1) mean that y tends to increase (or decrease) as x increases. As such, r is used to test for correlations between paired variables.

A word of warning about interpreting correlations: correlation is not the same as causation. Just because two observables are correlated they need not be causally connected to each other. For example: the number of grey hairs on my head and the number of mobile phones owned in the UK are correlated, not because one causes the other, but because they have both increased in time for quite different reasons.

R.Box 2.18
Computing the correlation coefficient

It is of course possible to compute the correlation coefficient r by computing each term in equation 2.6 separately and combining them, but the `cor()` function will do it for you. Or the `cor.test()` function will compute the coefficient and useful additional information. Using the fake data from R.box 2.14 we can use

```
cor(x, y)
cor.test(x, y)
```

This will display the correlation coefficient r (in the last line) and also information such as the p-value, to be discussed in later chapters.

2.9 Plotting multivariate data

So far we have discussed ways to present univariate and bivariate data. But many datasets are *multivariate*, that is, they contain information on many variables. These may contain information on many relationships between the variables, but they present more of a challenge to display on a two-dimensional screen or page.

One approach is to graph two variables as a scatter plot and use a change of symbols to represent a third variable. Symbol shapes can be discrete (e.g. circle, square, star) and work well for a discrete or categorical variable, whereas symbol sizes or shades are continuous and can represent a continuous variable. Figure 2.13 shows a *bubble plot*, which is related to the standard scatter diagram except that a third numerical variable is represented by a continuum of sizes for the symbols (for clarity we could also vary the colours/shades of the symbols). However, it is generally much harder to distinguish shades and areas than it is positions in two dimensions (see Cleveland, 1985), and so this method should not be relied upon to present detailed quantitative information on the third variable.

R.Box 2.19

Bubble plots

We shall demonstrate how to produce a bubble plot using some data on atmospheric conditions, described in Appendix B (section B.6). Using the four-column data array `env` (created in R.box B.13), we can produce a scatter plot of two variables (ozone concentration against temperature), and vary the symbol size (the `cex` argument) using a third variable (wind speed), as follows:

```
plot(env$temp, env$ozone, cex=env$wind/2)
```

We have scaled the wind speed variable by a factor of 1/2 to ensure the symbols are not too large or small. But still this could be improved by adding axis labels, increasing the size of the labels and text, defining the axis ranges, also adding some 'jitter' to the temperatures, and adding a smoothed curve. (The temperatures were stored only in integer °F units.)

```
plot(jitter(env$temp), env$ozone, cex=env$wind/2,
     xlab="Temperature (K)", ylim=c(0, 170),
     ylab="Ozone (ppb)", bty="l",
     cex.axis=1.2, cex.lab=1.4)
smooth <- lowess(env$temp, env$ozone, f=0.5)
lines(smooth, lwd=5, col="purple")
```

A legend can be added using the `legend()` function, as in Figure 2.13.

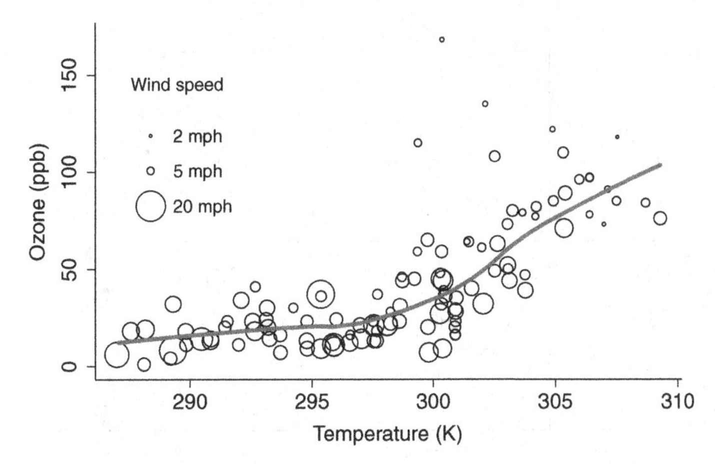

Figure 2.13 Example of a bubble plot. This is a scatter plot for two variables (ozone concentration and temperature), but the symbols vary in size to illustrate variations in a third variable (wind speed). A legend on the left gives some indication of the scale of the third variable. Hollow circles (bubbles) have been used here to help distinguish symbols that partially overlap. A smoothed curve has also been plotted to emphasise the non-linear correlation between ozone concentration and temperature.

An alternative approach is to use a multipanel plot combining several different scatter plots. One example of this is the *coplot* (*co*nditioning *plot*) (Cleveland, 1993), which is made from a series of scatter plots for two variables, where each separate panel uses only a subset of the data selected using the value of a third variable. Another very powerful way to visualise a multivariate dataset is by plotting one scatter plot for every possible pair of variables and combining these in a matrix, as in Figure 2.14. Cleveland (1993) called this 'one of the best visualization ideas around'. The aficionado of statistical graphics may not like the redundancy in this plot: the top-right panels repeat the same information as the bottom-left panels. Even more information can be added to such a plot by replacing one set of panels (e.g. top right) by alternative presentations, for example contour density plots (see below) or numerical values of the correlation coefficients for each pair of variables.

R.Box 2.20
Matrix of scatter plots

The simplest way to produce a scatter plot matrix in R is with the `pairs()` command. The following commands produce a matrix plot of the atmospheric data `env` (see Appendix B, section B.6), like that of Figure 2.14:

```
pairs(env)
```

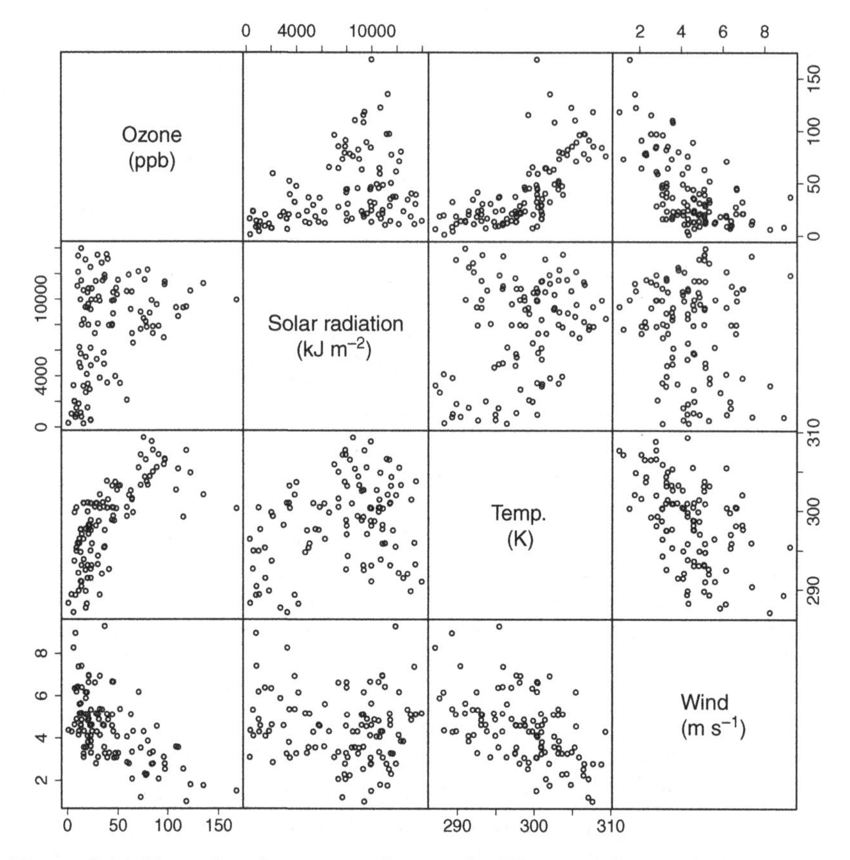

Figure 2.14 Example of a scatter plot matrix. There are four variables – ozone concentration, solar radiation, temperature and wind speed – recorded on each of 111 days in 1973 in the New York City metropolitan region. The data can be considered as 111 points in four-dimensional space. This graphic shows all the different possible pairs of variables in separate scatter plots arranged in a matrix with shared axes.

See `?pairs` for more details. We can improve the clarity of the plot and reduce wasted space by defining more informative text labels by setting some options:

```
labels <- c("Ozone\n(ppb)",
            "Solar radiation\n(kJ/m^2)",
            "Temp.\n(deg F)", "Wind\n(m/s)")
pairs(env, labels=labels, cex.labels=1.5, gap=0)
```

The first line defines a list of four labels for the four variables that includes units. (By default the `pairs()` command uses the names of the data columns as labels.) The

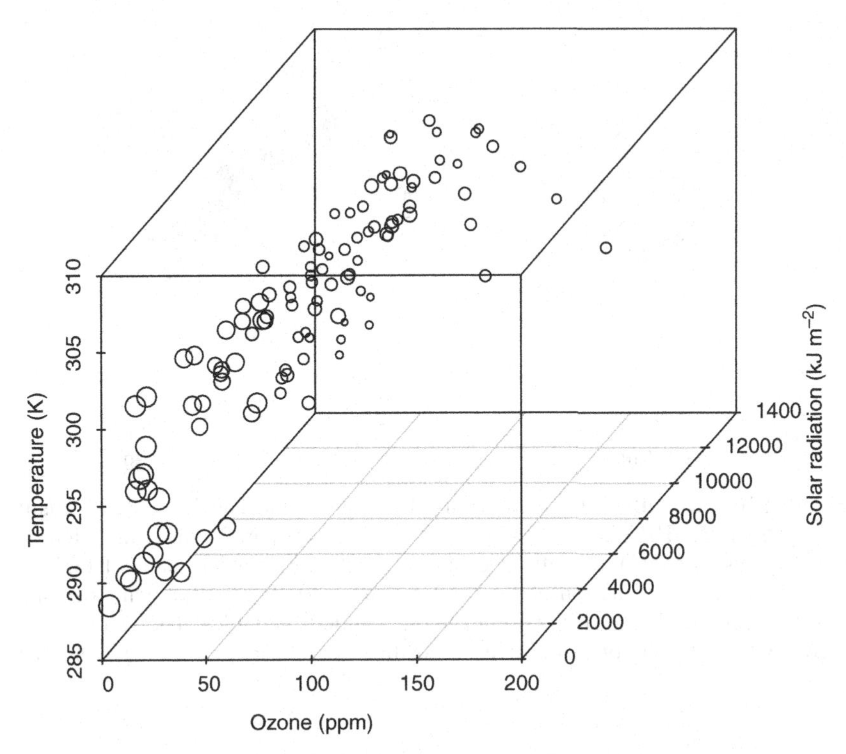

Figure 2.15 Projected three-dimensional scatter plot. The data show the atmospheric measurements of Figures 2.13 and 2.14. The 'depth' dimension (solar radiation) can be particularly difficult to distinguish, but this could be improved by shading/colouring the symbols to represent one of the variables. Here we use the symbol size to emphasise the position along this dimension. (Created in R using the `scatterplot3d` package.)

second line calls the `pairs()` command, but this time we have specified the labels and the text size (`cex.labels=1.5`), and removed the gaps between individual panels (`gap=0`).

2.9.1 Three-dimensional projections

Figure 2.15 shows an example of a scatter plot with three variables. This is an attempt to project an intrinsically three-dimensional structure onto a two-dimensional plot. Projections like this can be a useful tool for exploratory data analysis, especially when animated or used interactively.

Figures 2.16 and 2.17 show three different ways to represent the same data. These are based on the HR diagram of Figure 2.6. The scatter plot is made from a sample of points, but we can form a two-dimensional function by estimating the density

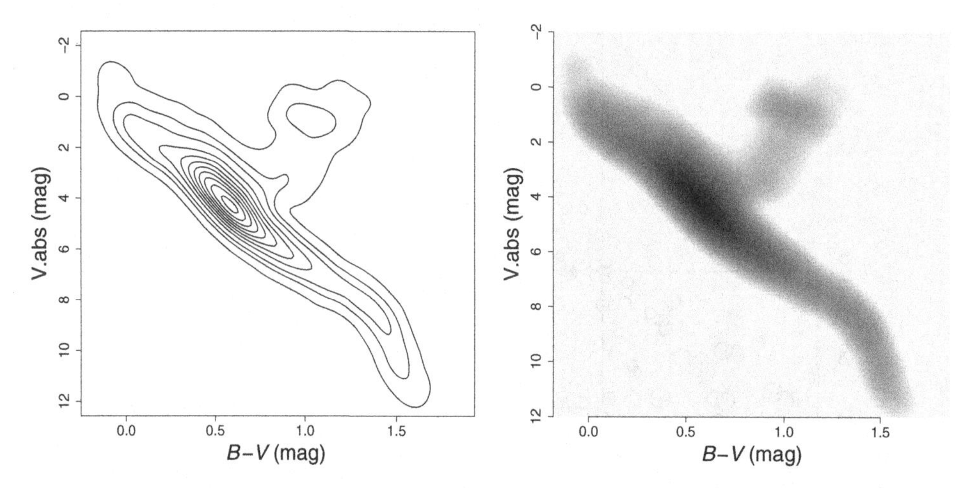

Figure 2.16 Two different visualisations of the same data. The data are the density of points on the HR diagram of Figure 2.6. The right panel shows an 'intensity map' (or just 'image'); the left panel shows only contours of the same data. Both allow us to view a function of the form $z = f(x, y)$, or measurements of some independent variable as a function of two dependent variables. The value of z is given by the intensity or colour at each position on an even grid of x, y positions.

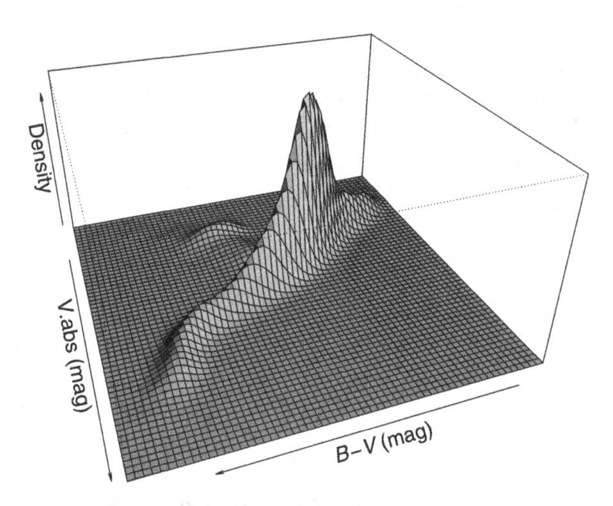

Figure 2.17 Same data as shown in Figure 2.16 but in the form of a surface projection plot. Notice that the axes have been rotated in order to reveal more structure.

of points over the plane. This is just like any function $z = f(x, y)$; in this case z is the density of points computed on a grid of x, y values (actually $B - V$ and V_{abs} values in this example). These alternative graphic representations of a surface tend to draw the viewer's attention to different aspects of the data. They can also be combined: for example, we could overlay contours on a colour image. This is often used in astronomy to compare different images of the same part of the sky, e.g. radio image contours over the optical image intensity map.

R.Box 2.21

Intensity, contour and three-dimensional projection maps

Given a two-dimensional array of numbers it is simple to produce an intensity image, a contour map or a surface projection using the `image()`, `contour()` and `persp()` commands. First, let's take the star data (Appendix B, section B.4) and compute the density of data points as a function of the two variables $B - V$ and V_{abs} using the `kde2d()` function (part of the `MASS` package):

```
require(MASS)
hr.dens <- kde2d(hip$BV, hip$V.abs, n=150)
image(hr.dens, col = terrain.colors(100),
      ylim=c(12,-2), xlab="B-V (mag)",
      ylab="V.abs (mag)")
```

The object `hr.dens` contains three variables: `hr.dens$x` and `hr.dens$y` are vectors of length 150 spanning the $B - V$ and V_{abs} dimensions, respectively, and `hr.dens$z` is a two-dimensional array (150×150) containing a smooth estimate of the density of points at a given position on the grid of $B - V$, V_{abs} values.

The equivalent contour and projection plots may be generated as follows:

```
persp(hr.dens, theta=160, phi=30,
         shade=0.5, col="grey", expand=0.5)
levels <- c(0.01, 0.05, seq(0.1,1,by=0.1))
contour(hr.dens, levels=levels,
        ylim=c(12,0))
```

The `levels` vector is a list of the densities at which to draw each contour. These commands are quite powerful and there are many options you may wish to change to customise the graphics.

2.10 Good practice in statistical graphics

There are some basic principles that are worth considering when trying to make clear and informative statistical graphics. Data graphics are used to reveal to the viewer as much as possible about the data, or to illustrate an idea clearly. You invest the hard work into producing the clearest and most informative visualisation so that the viewer needs to invest relatively little work in understanding it.

The list below gives some of these guidelines for producing good data graphics, based loosely on the recommendations of Tufte (1986) and others.

- Show your data, without distortion, and with clear labelling (and/or a clear caption).
- Use a graphic appropriate to the data or idea.
- Try to show the greatest amount of information as clearly as possible in the available space.
- Use colour carefully (and be considerate to colour-blind viewers). Don't expect the viewer to distinguish between seven shades of green!
- If a single plot does not convey all the information in a clear fashion then use multipanel plots (e.g. two panels side by side, or a 3×3 grid).

2.11 Chapter summary

- Sample mean for a sample of n observations x_i ($i = 1, 2, \ldots, n$)

$$\bar{x} = \frac{1}{n}\sum_{i=1}^{n} x_i$$

- The median is the middle point of a dataset (the 50th percentile, or 0.5 quantile)
- The mode is the most popular value within a dataset (or the peak of a density distribution)
- Sample variance, s_x^2, and standard deviation, s_x

$$s_x^2 = \frac{1}{n-1}\sum_{i=1}^{n}(x_i - \bar{x})^2 \qquad s_x = \sqrt{s_x^2}$$

- Standard error (on sample mean)

$$\mathrm{SE}_{\bar{x}} = \sqrt{s_x^2/n}$$

- The sample correlation coefficient r, for a dataset comprising n observations of x_i, y_i, is

$$r = \frac{s_{xy}}{s_x s_y}$$

where s_{xy} is the sample covariance

$$s_{xy} = \frac{1}{n-1}\sum_{i=1}^{n}(x_i - \bar{x})(y_i - \bar{y})$$

- Plotting distributions
  - Histogram – for binned continuous data
  - Bar chart – for distribution of discrete/categorical data
  - Dot chart – shows continuous against categorical data
  - Box plot – summary of one or more distributions side by side
  - Rug – for augmenting a histogram (or scatter plot) with actual data values
- Scatter plot – for showing y against x
  - Choose simple symbols that stand out against the background
  - Define the axis ranges to include all (or most) of the data without wasting too much empty space
  - Use different symbols/colours to distinguish a categorical variable (e.g. different types of subject or experimental set-up)
  - Plot the variables in a way that reveals the most about the data (e.g. plot y, or $\log y$, or $y - x$)
- Functions of two variables may be plotted using
  - Surface projection
  - Intensity/colour map images
  - Contour plots
- Scatter plots with more than two variables may be represented in various ways:
  - Use symbol size/shade to represent third variable
  - Three-dimensional projection of scatter plot (for three variables only)
  - Multipanel scatter plots (scatter plot matrix and coplot)

3

Simple statistical inferences

Everything should be made as simple as possible, but not simpler.

Attributed to Einstein

We can use what we have learnt to start making some inferences about data. Maybe we have collected measurements of a quantity and wish to see if these are consistent with some theoretical expectation. We don't just want to compute the sample mean but to compare it with something else. Perhaps we have two samples, taken under different conditions (such as a 'treatment' and 'control' group) and wish to see if their mean responses differ. Another very common situation is that we have measurements of some response (y) taken at different values of some explanatory variable (x) and wish to quantify the way that y responds. We can go some way to getting useful inferences out of such data using numerical and graphical summaries (Chapter 2). These can be refined once we have studied some probability theory (Chapters 4 and 5).

3.1 Inference about the mean of a sample

We take repeated measurements of a single quantity, or measure the same quantity for each member of a finite sample, and wish to discover whether these data are consistent with a predetermined theoretical value. We want to know if our sample is consistent with being randomly drawn from a theoretical population, with some particular population mean. As an example, let's consider the first 'experiment' (batch of 20 runs) of Michelson's dataset (see Appendix B, section B.1). This comprises 20 of his speed of light measurements taken under similar experimental conditions:

850	740	900	1070	930	850	950	980	980	880
1000	980	930	650	760	810	1000	1000	960	960

How should we compare these to the modern value[1] of 734.5 km s^{-1}? Let's call these data x_i $(i = 1, 2, \ldots, 20)$. The simplest way forward is to compute the sample mean, in this case $\bar{x} = 909$ km s^{-1}, and compare this to our predicted value $\mu = 734.5$ km s^{-1}, by taking the difference: $\bar{x} - \mu = 174.5$ km s^{-1}. This is the offset between the 'centre of mass' of our small data sample and our predicted value.

But in order to put this in context, we need to know how this compares to the expected spread of the data. In particular, is this difference comparable to the standard error, or is it larger? If it is comparable, then it seems plausible that the observed difference $\bar{x} - \mu$ is due only to the finite precision of $\bar{x}$, which in turn is limited by the size of the data sample ($n = 20$) and the spread within the sample (presumably due to random 'experimental error'). We can quantify this in terms of the ratio of the difference (between sample mean and expect values) to the standard error on the mean (equation 2.4). The result is a dimensionless statistic:

$$t = \frac{\text{observed difference}}{\text{standard error}} = \frac{\bar{x} - \mu}{\sqrt{s_x^2/n}}. \tag{3.1}$$

(The standard error is defined so long as $n \geq 2$, but really small samples, e.g. $n \lesssim 10$, should be treated very carefully.) This is known as Student's t-statistic.[2] In fact, there are many variations on the t-statistic; this one is used for single samples.

R.Box 3.1

Computing the one-sample t-statistic

It is possible to compute the t-statistic with a single command in R, but we begin by explicitly computing the statistic from its components. First, we define our data sample, in this case the first 'experiment' from the Michelson data.

```
subset.1 <- (morley$Expt == 1)
x <- morley$Speed[subset.1]
```

Then we can compute the sample mean, define the predicted mean value, compute the sample mean standard error and combine these to form the t-statistic.

[1] This number was computed by Stigler (1977) by correcting the modern value for the speed of light in a vacuum (299 792.5 km s^{-1}) using Michelson's corrections to the speed of light through air. Remember, the data have had 299 000 km s^{-1} subtracted.

[2] 'Student' was the pen-name of William Sealy Gosset, a talented statistician who derived many interesting results, including the t-statistic (in 1908), while working for the Guinness Brewing Company. Company policy prohibited publication, so Gosset published under the pseudonym 'Student'.

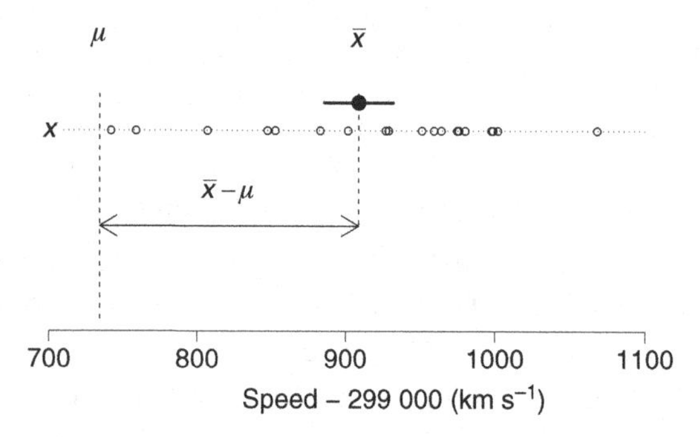

Figure 3.1 Illustration of the t-statistic for the difference of a mean. The points represent the $n = 20$ data points from the first 'experiment' of the Michelson data (with some 'jitter' to help distinguish overlapping points). The large black point shows the sample mean, and the error bars show the standard error on the mean. The vertical dashed lines show the position of the sample mean $\bar{x}$, the predicted mean μ and their difference. The t-statistic is the ratio of this difference to the standard error.

```
mean.x <- mean(x)
mu <- 734.5
se.x <- sqrt(var(x) / length(x))
t.stat <- (mean.x - mu) / se.x
```

You should find the variable `t.stat` has the value 7.438.

R.Box 3.2

The one-sample t-statistic with a single function

We can compute the statistic using the R function `t.test()` as follows:

```
t.test(x, mu=734.5)
```

The inputs are the data x, and the predicted mean value is specified with the `mu` argument. The output includes the t-statistic value, but also some information about the significance test and confidence interval, which are the subjects of Chapter 7.

Figure 3.1 illustrates the t-statistic using the first 20 of the Michelson data values. In this case we have a standard error of 23.46 km s^{-1} (see section 2.6) and so

$$t = \frac{174.5}{23.46} = 7.438. \tag{3.2}$$

In words, the sample mean differs from the predicted value by more than seven standard errors. This is also shown in Figure 2.5. Physicists often use 'sigma' (σ) as shorthand for the standard error, and would say this is a '7σ' difference. Is this a lot?

One standard error is an estimate of the typical fluctuation of the sample mean about the true mean. Seven standard errors means a larger than typical fluctuation from the predicted value, if the predicted value is correct. Alternatively, the predicted value is wrong, or the standard error is too small. Once we have studied more probability theory we shall return to these ideas and establish a more quantitative test.

3.2 Difference in means from two samples

A related problem is comparing the means between two samples. Here we have two samples of data, perhaps taken under different experimental conditions, and we are not interested in whether the means differ from some theoretical value but whether they differ between samples. This is one of the most widely used experimental procedures and tests in all of science: the comparison of the mean response under two different treatment conditions. For example, the efficacy of a medical intervention can be tested by comparing the mean responses between a 'treatment' group and a 'control' group (without the treatment). Or, the effect of a new particle background removal system can be examined by comparing the mean particle background with and without the system in operation.

Let's return to the Michelson dataset, and compare the results of the first two 'experiments', that is the first two batches of $n = 20$ 'runs'. The second experiment yielded the following data values; let's call these data y_i $(i = 1, 2, \ldots, 20)$:

960	940	960	940	880	800	850	880	900	840
830	790	810	880	880	830	800	790	760	800

The comparison of two samples is based on a generalisation of the t-statistic. We first compute the means of each sample, $\bar{x}$ and $\bar{y}$, and their respective standard errors $\sqrt{s_x^2/n}$ and $\sqrt{s_y^2/n}$, and then form a t-statistic of the form

$$t = \frac{\text{difference of means}}{\text{standard error (of difference of means)}} = \frac{\bar{x} - \bar{y}}{\sqrt{s_x^2/n + s_y^2/n}}. \tag{3.3}$$

The denominator is the standard error on the difference, formed from the square root of the sum of the individual squared standard errors. (Variances combine linearly, as we shall discuss in Chapter 5.)

The above is an example of the two-sample t-statistic, assuming the two samples are of equal size ($n_x = n_y = n$) and of equal population variance. We can look again at the one-sample t-statistic (equation 3.1) and see it as a special case of the two-sample statistic for which the mean of the second sample is known perfectly,

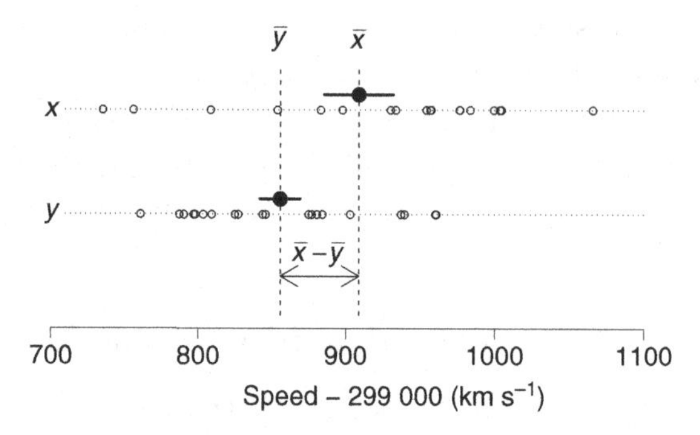

Figure 3.2 Illustration of the t-statistic for the difference between two means. The upper points represent the $n = 20$ data points from the first 'experiment' of the Michelson data, x, and the lower points represent the $n = 20$ data points from the second 'experiment'. The large black points show the sample means, and the error bars show their standard errors. The two-sample t-statistic is formed from the difference between the means, as a ratio to the standard error of the difference.

that is there is zero standard error on the second sample mean $\mathrm{SE}_y = 0$. There are further generalisations of the t-statistic to uneven sample sizes, to unequal variances for the two samples and even to more than two samples. The formulae for these t-statistics are more complicated, but the basic idea is the same.

R.Box 3.3
Computing the two-sample t-statistic

As before, we begin by explicitly computing the statistic from its components. First, we define our second data sample, the second 'experiment' from the Michelson data.

```
subset.2 <- (morley$Expt == 2)
y <- morley$Speed[subset.2]
```

Then we can compute the two-sample t-statistic in steps

```
mean.y <- mean(y)
se.y <- sqrt(var(y) / length(y))
t.stat <- (mean.x - mean.y) / sqrt(se.x^2 + se.y^2)
```

You should find that `t.stat` has the value 1.952.
We can compute the two-sample statistic using the `t.test()` as follows:

```
t.test(x, y, var.equal=TRUE)
```

The inputs are the first and second data samples, and we have also specified that we expect the populations variances to be the same. Again the output includes information that can help us interpret the t-statistic.

Table 3.1 *Example dataset: data for response variable y taken at different values of the explanatory variable x.*

x	10.0	12.2	14.4	16.7	18.9	21.1	23.3	25.6	27.8	30.0
y	12.6	17.5	19.8	17.0	19.7	20.6	23.9	28.9	26.0	30.6

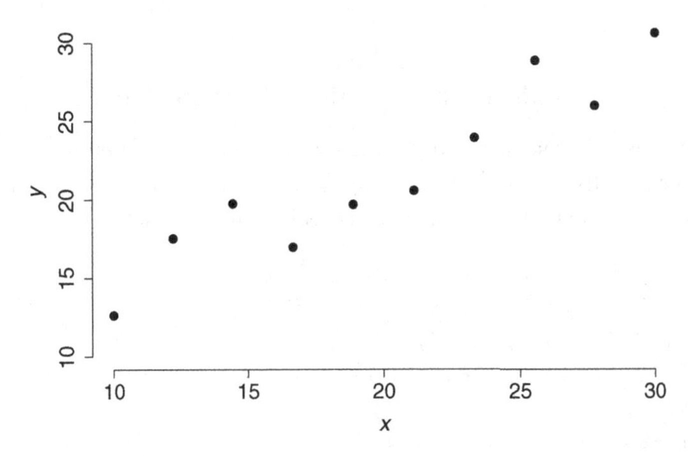

Figure 3.3 A scatter plot showing data for response variable y taken at different values of the explanatory variable x. How does y change with x?

Returning to the simple two-sample case, for the Michelson data we can easily find $\bar{x} = 909$ and $\bar{y} = 856$; also, the standard errors are $\sqrt{s_x^2/n} = 23.5$ and $\sqrt{s_y^2/n} = 13.7$ (all of these have units of km s^{-1}). And so

$$t = \frac{909 - 856}{\sqrt{23.5^2 + 13.7^2}} = 2.0. \tag{3.4}$$

This tells us that the means differ, but the magnitude of the difference is only twice the standard error of the difference between two means from samples like these. Figure 3.2 illustrates the two-sample t-statistic using the first two 'experiments' of the Michelson data.

3.3 Straight line fits

The t-statistic provides a basis for comparing two samples. In effect we have bivariate data; each data point comprises a number (the response measurement) and a categorical variable (the explanatory variable) that can take on two values,[3] 'group 1' or 'group 2'. We have looked at how to quantify any changes in the mean

[3] In the statistics jargon, the categorical variable is called a *factor*, and the different values it can take are called *levels*.

between the groups, i.e. how the response changes as the explanatory variable changes. Now, how do we quantify the change of a response variable with a continuous explanatory variable?

Table 3.1 shows some example data, and Figure 3.3 displays them. The question is: how does y change with x?

R.Box 3.4

Simple linear regression – example data

The following is a demonstration of simple linear regression, computing the coefficients explicitly, using a dataset comprising $n = 10$ (x, y) points. We first input the data, which we do manually since this is such a small dataset.

```
x <- c(10.0, 12.2, 14.4, 16.7, 18.9,
       21.1, 23.3, 25.6, 27.8, 30.0)
y <- c(12.6, 17.5, 19.8, 17.0, 19.7,
       20.6, 23.9, 28.9, 26.0, 30.6)
plot(x, y, bty="n", pch=16,
   xlim=c(10, 30), ylim=c(10,30))
```

Of course, the first thing we do with the data is make a plot. This immediately allows us to check that we have inputted the correct data, and look for patterns and outliers.

We shall restrict ourselves to linear models here, i.e. a straight line relationship between response and explanatory variables.

$$y = f(x) = \alpha + \beta x \tag{3.5}$$

for some specific values of the coefficients α and β. The model should predict the response variable, y, as a function of the explanatory variable, x. Of course, any realistic data will be subject to some experimental error. We should therefore not expect the linear model, with the correct coefficients $\alpha = \alpha_0$ and $\beta = \beta_0$, to predict a real dataset perfectly, but to predict the (error free) population mean of y as a function of x. We can summarise all this by saying

$$y_i = (\alpha_0 + \beta_0 x_i) + \varepsilon_i. \tag{3.6}$$

Here, y_i are the response data ($i = 1, 2, \ldots, n$), x_i are the explanatory data and ε_i are the random 'errors'. Also, α_0 and β_0 are the true values of the coefficients (intercept and gradient) of the linear function. If we knew α_0 and β_0, and knew how the ε_i were distributed, we would know everything about the data generation process, a full statistical model. But we do not know α_0 and β_0; all we know is some particular values of x_i, y_i, and we wish to find some reasonable estimates of the coefficients.

3.3.1 Fitting by least squares

If we have a model in mind we can compare the observed y_i values with those predicted using the model. If we choose our model well, then the predicted and observed values should be reasonably close, at least within the limits set by the experimental errors (ε_i). The difference between the predicted and observed mean values is called the *residual*. We can write this schematically:

$$\text{data} = \text{model} + \text{residual}. \tag{3.7}$$

We could guess some values for the coefficients of the model (3.6), and look at the residuals to see how well the predictions match the data:

$$\begin{aligned} \text{residual} &= \text{data} - \text{model} \\ e_i &= y_i - (\alpha + \beta x_i). \end{aligned} \tag{3.8}$$

Notice that we are being careful to distinguish between the errors ε_i (differences between data and 'true' model) and the residuals e_i (differences between data and our chosen or estimated model). If we guessed the correct values for the coefficients, then the residuals would be the experimental errors $e_i = \varepsilon_i$. If our guesses were badly off, we would expect the residuals to be larger in magnitude, on average. What we need is a system for finding the coefficients that give reasonably small residuals.

The simplest and most popular solution to this problem is to find the coefficients of the model such that the sum of the squared residuals is at a minimum: in other words, find where

$$\text{SSE} = \sum_{i=1}^{n} e_i^2 = \sum_{i=1}^{n} [y_i - (\alpha + \beta x_i)]^2 \tag{3.9}$$

is minimised. Why do we square the residuals? The simplest answer is that the squared residuals are non-negative, so positive and negative residuals do not cancel out. In Chapter 6 we shall give a more satisfying answer. You might also have noticed that the SSE statistic defined here bears a strong resemblance to the sample variance (e.g. equation 2.3).

Figure 3.4 shows the solution for our example dataset. The solid line shows the best-fitting linear model, and the residuals are shown as vertical lines connecting

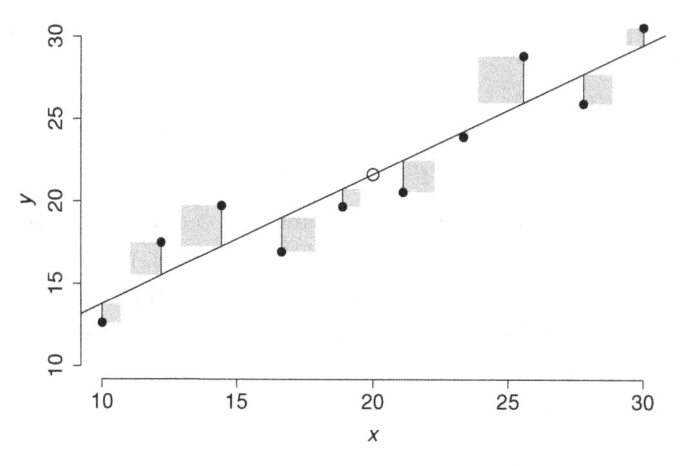

Figure 3.4 Illustration of simple linear regression. The data points are the same as in Figure 3.3, and the hollow circle shows the data 'centre' $(\bar{x}, \bar{y})$. The solid line shows the simple linear regression model through the data. The vertical lines (connecting data and model) show the residuals, and the grey squares illustrate the squared residuals. The total area in the grey squares represents the sum of the squared residuals. The line is the best-fitting line in the sense that any other line would give a larger sum squared residual, i.e. larger total area of grey boxes.

the data points to the model prediction at each x_i value. The grey squares illustrate the square residuals; the sum of the squared residuals is equal to the total area of these squares, and the best-fitting model is the one that minimises this area.

3.3.2 Finding the solution

We can find the values of α and β that minimise SSE using standard mathematical tools, for example differentiate SSE and find where its gradient is equal to zero. (The reader not interested in the derivation may wish to skip this section.) We find a and b such that

$$\left.\frac{\partial S}{\partial \alpha}\right|_{\alpha=a} = 0 \quad \text{and} \quad \left.\frac{\partial S}{\partial \beta}\right|_{\beta=b} = 0. \tag{3.10}$$

Now, inserting the derivatives, we find a and b from the pair of simultaneous equations

$$-2\sum_{i=1}^{n} y_i - a - bx_i = 0$$

$$-2\sum_{i=1}^{n} x_i\,[y_i - a - bx_i] = 0. \tag{3.11}$$

R.Box 3.5

Simple linear regression – the long way

Next, we compute the various quantities that are needed to estimate the intercept and gradient of the linear model

```
mean.x <- mean(x)
mean.y <- mean(y)
mean.x2 <- mean(x^2)
mean.xy <- mean(x*y)
b <- (mean.xy - mean.x*mean.y) /
      (mean.x2 - mean.x^2)
a <- mean.y - b * mean.x
abline(a, b, lwd=2, col="red")
```

The last line adds a straight line to the plot with the intercept and gradient as computed from the linear regression solution.

These can be simplified by dividing both sides by $-2n$:

$$\begin{aligned} \bar{y} - a - b\bar{x} &= 0 \\ \overline{xy} - a\bar{x} - b\overline{x^2} &= 0, \end{aligned} \tag{3.12}$$

where we have made the substitutions

$$\bar{x} = \frac{1}{n}\sum_{i=1}^{n} x_i, \quad \bar{y} = \frac{1}{n}\sum_{i=1}^{n} y_i, \quad \overline{xy} = \frac{1}{n}\sum_{i=1}^{n} x_i y_i, \quad \overline{x^2} = \frac{1}{n}\sum_{i=1}^{n} x_i^2. \tag{3.13}$$

After a little more algebra, we get the solution

$$\begin{aligned} b &= \frac{\overline{xy} - \bar{x}\bar{y}}{\overline{x^2} - \bar{x}^2} \\ a &= \bar{y} - b\bar{x}. \end{aligned} \tag{3.14}$$

This means that given some data pairs (x_i, y_i) we can find the gradient b and intercept a of the straight line that will minimise the sum of the squared residuals between the data y_i and the predictions of the model. The predicted data values are then

$$\hat{y}_i = a + bx_i. \tag{3.15}$$

By expanding and rearranging, it is possible to write the expression for b in several different ways,

$$b = \frac{\frac{1}{n}\sum_{i=1}^{n}(x_i - \bar{x})(y_i - \bar{y})}{\frac{1}{n}\sum_{i=1}^{n}(x_i - \bar{x})^2} \tag{3.16}$$

which can be compared with the definitions of the sample covariance s_{xy} and correlation coefficient r (equations 2.5 and 2.6) to find

$$b = \frac{s_{xy}}{s_x^2} = r\frac{s_y}{s_x}. \tag{3.17}$$

3.4 Linear regression in practice

The process of finding the coefficients of a model that minimise the sum of the squared residuals is often called *linear regression* or *least-squares fitting* – the former term is more popular among statisticians, life and social scientists; the latter term is more popular among physical scientists. By following equations 3.14, we can find the regression (or least-squares) estimates of the coefficients α and β. At least for small datasets this can be done manually, but in practice we would almost always use a computer to perform the calculations (see R.box 3.5). But most good data analysis software, including R, will do linear regression for you with a single command or function (see R.box 3.6).

The results The result of a simple linear regression analysis is a pair of coefficients (the intercept and gradient). It is also possible to compute standard errors on the coefficients – but we shall reserve discussion of this until Chapter 7. The estimated coefficients specify a linear function that passes through the 'centre' of the data $(\bar{x}, \bar{y})$. Also, the sum of the residuals will cancel perfectly, i.e. $\sum e_i = 0$. (We leave it as a exercise for the student to prove these relations.)

Assumptions We should be careful to examine the assumptions implicit in this kind of analysis. First, we have minimised the squared residuals with respect to the y data, taking no account of any uncertainty on the x data. In other words, we have assumed that the uncertainties on the data are dominated by those on the response variable, and the uncertainty on the explanatory variable can be neglected. This is often the case if the explanatory variable is something that can be tightly controlled during the experiment. We have also assumed that the random (experimental) errors ε_i are not correlated with each other, and have zero mean and equal weight. In other words, the y_i data are randomly scattered around the 'true' model $f(x_i)$, and there is no structure to the errors.

Linearity The model used above is linear, but often the relationship between variables is not expected to be linear. Even so, simple linear regression can often be used after applying a simple transformation to the data. For example, if we have data x_i, y_i and wish to fit a model of the form $y = \alpha x^{\beta}$, we could plot and fit $\log y_i$ against $\log x_i$, transforming the power law relationship into a linear one.

Other models The method outlined above, using a model that is linear in x, is called *simple* linear regression (simple because the model is simple). Linear regression is more general in that the model can be extended to be non-linear in x, but still linear in its coefficients, for example

$$f(x) = \alpha + \beta x + \gamma x^2 + \delta \log x. \quad (3.18)$$

This is linear in its coefficients ($\alpha, \beta, \gamma, \delta$), but non-linear in x. It is possible to determine the least-squares estimates for such models in a manner similar to that shown above, but in such cases we formulate the problem using the tools of linear algebra and then perform the calculation on a computer.

Connection to t-tests There is a close relationship between linear regression and the comparison of two sample means discussed above (section 3.2). One can think of comparing the means between two samples as a regression analysis on the data (x_i, y_i), where $x_i = 0$ or 1 depending on which group the y_i belongs to. In this case the gradient is equal to the difference in the means.

R.Box 3.6

Simple linear regression – the short way

R features a powerful array of functions for regression. We can perform simple linear regression using a single function

```
result <- lm(y ~ x)
```

where the `lm()` function (short for *linear model*) works to find 'linear models' by regression, and it takes as input a 'formula' relating variables. In this case we specify to regress y on x, i.e. find a model that predicts y with a linear function of x. The output of this function is stored in our variable `result`, which contains a lot of information.

```
summary(result)
```

The part we are interested in here is `result$coefficients`, which is a two-element list of the model coefficients (intercept and gradient). These should match those we calculated previously (R.box 3.5). There are many functions that can make use of the output of the `lm()` function. For example, we could overlay a plot of the linear fit like this:

```
abline(result, lwd=2, col="blue")
```

3.5 Residuals: what lies beneath

After performing a regression analysis it is important to check the model fit. We can do this using graphical and numerical summaries. The simplest thing to do is to plot the data together with the model. Depending on the data, it may be more revealing to plot the (data–model) residuals – structure in the residuals could be a sign of structure in the data that is not properly accounted for by the model. It is also important to check for outliers – a single extreme value of x and/or y can have a powerful influence on the results.

We can also form numerical breakdowns of the regression. Look again at equation 3.7. We can express this as

$$(y_i - \bar{y}) = (\hat{y}_i - \bar{y}) + (y_i - \hat{y}_i) \tag{3.19}$$

where y_i are the data, $\bar{y}$ is the sample mean of the data and $\hat{y}_i$ are the values predicted from the regression model. The left side represents the data with its sample mean subtracted. The two terms on the right side represent the model (predicted) values also with the sample mean subtracted, and the residuals. Now it is also true that the sums of the squares of these three deviations also add:

$$\sum_{i=1}^{n}(y_i - \bar{y})^2 = \sum_{i=1}^{n}(\hat{y}_i - \bar{y})^2 + \sum_{i=1}^{n}(y_i - \hat{y}_i)^2. \tag{3.20}$$

This may not be immediately obvious. The proof is rather long winded and involves squaring and summing both sides of equation 3.19, systematically expanding the terms and making use of the fact that $\sum e_i = 0$ to cancel terms (where $e_i = y_i - \hat{y}_i$).

Each of these terms is a sum of squares and has its own name

$$\mathrm{SST} = \sum_{i=1}^{n}(y_i - \bar{y})^2, \quad \mathrm{SSM} = \sum_{i=1}^{n}(\hat{y}_i - \bar{y})^2, \quad \mathrm{SSE} = \sum_{i=1}^{n}(y_i - \hat{y}_i)^2. \tag{3.21}$$

Here SST is the sum of the squared *total* deviations, SSM is the sum of the squared *model* deviations and SSE is the sum of the squared *error* values.

Each of these sums of squares is like a sample variance, except for a factor $1/(n-1)$. SST represents the total variance in the y data, SSM represents the variance in the predicted values, that is the variance in the data we can explain in terms of the model, and SSE represents the residual or 'unexplained' variance. This is known as *partitioning* the sums of squares.[4] These three numbers tell us how much of the total variance in our response is accounted for by our linear

[4] If the experimental errors are known, it is possible to further partition SSE into the variance due to the experimental errors and the variance due to the poorness of the model fit.

model and how much remains unaccounted for by the model. By construction the regression analysis gives us the coefficients for the model that minimise SSE, which means they maximise SSM.

The SSE is the 'unexplained' variance, which we might initially assume was due to the 'errors'. We can therefore use SSE to estimate the size of the errors, that is the variance of the random ε in equation 3.6,

$$S^2 = \frac{\text{SSE}}{n-2} = \frac{1}{n-2}\sum_{i=1}^{n}(y_i - \hat{y}_i)^2. \tag{3.22}$$

This is very much like the sample variance of the data around the model prediction (compare with 2.3), except that now the sum is multiplied by a term $1/(n-2)$. The $n-2$ is used because the model has two coefficients that were estimated from the data, chosen because they minimised the square residuals, before calculating the residual variance. Without the -2 correction the estimate would be (on average) too low, because it is based on the model parameters selected to give the smallest possible residual variance.

Now, we can consider the ratio

$$r^2 = \frac{\text{SSM}}{\text{SST}} = \frac{\sum_{i=1}^{n}(\hat{y}_i - \bar{y})^2}{\sum_{i=1}^{n}(y_i - \bar{y})^2}. \tag{3.23}$$

This quantifies the fraction of the variance of y that is 'explained' by the model. This is the same as the square of the linear correlation coefficient r (see section 2.8).

3.6 Case study: regression of Reynolds' data

Appendix B, section B.3, describes data taken by Osborne Reynolds on the flow of a fluid (water) through a pipe. The data table gives his records of the average fluid velocity, v, at various values of the pressure gradient, $\Delta P/\Delta L$. We can learn a lot about these data using simple linear regression.

A linear fit to all the data shows a problem: there is structure left in the residuals. Specifically, the residuals rise to a peak at a pressure gradient of ≈ 65 Pa m^{-1}, and then fall off again. Reynolds noticed that at a pressure gradient of 70 Pa m^{-1} or higher the flow became turbulent ('unsteady'). In order to study only the laminar (streamline) flow, we can ignore these data, that is include only the first eight data points, and perform another linear regression. The result is shown in Figure 3.5. The numerical and graphical summaries (e.g. plots of residuals, and $r^2 = \text{SSM/SST}$) suggest that this model gives a good match to the data.

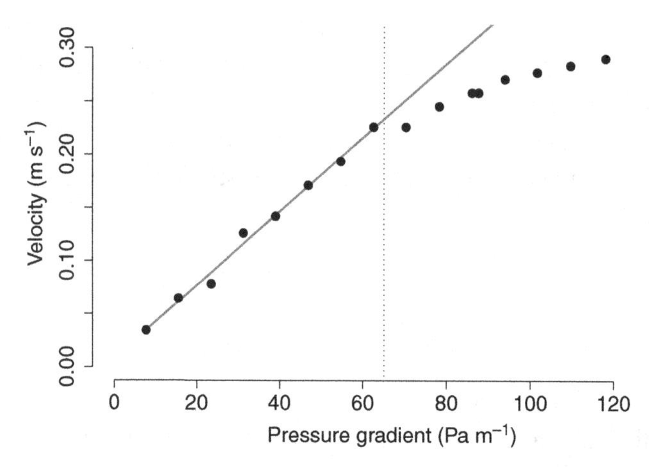

Figure 3.5 Reynolds' fluid flow data, compared with a simple linear model (a regression line) fitted only to the first eight data points, that is at pressure gradient ≤ 65 Pa m^{-1} (dotted line). The model matches the data extremely well until this transition point, above which the velocity increases more slowly with increasing pressure gradient.

R.Box 3.7

Linear regression of Reynolds' data

We first load Reynolds' data from a file (see Appendix B, section B.3), then perform simple linear regression (using the `lm()` function) on the variables `v` and `dP`.

```
fluid <- read.table("fluid.txt", header=TRUE)
result <- lm(v ~ dP, data = fluid)
```

We can examine the results using `summary()`, and plots, for example of the data with fitted model, and of the data–model residuals.

```
plot(fluid$dP, fluid$v, bty="n", cex=1.5, pch=16,
     ylab="Velocity (m/s)",
     xlab="Pressure grad (Pa/m)",
     xlim=c(0,125), ylim=c(0,0.3))
summary(result)
abline(result, lwd=2, col="red")
plot(fluid$dP, result$resid, pch=16)
```

It should be clear that something is not quite right here.

The gradient of this relation is grad $= 3.47 \pm 0.12 \times 10^{-3}$, that is the relative error is $\approx 3.5\%$ (for now we shall simply accept the standard error reported from the regression analysis). Poiseuille's equation says that this gradient should be $R^2/8\eta$

(e.g. equation B.3), where R is the pipe radius and η is the dynamic viscosity. In this experiment $R = 6.35 \times 10^{-3}$ m, so we can estimate the viscosity from $R^2/(8 \times \text{grad}) = 1.45 \times 10^{-3}$. We can give an approximate standard error as 3.5% of this, that is our estimate is $\hat{\eta} = 1.45 \pm 0.05 \times 10^{-3}$ Pa s. This compares to the modern value for the viscosity of water (at 8 °C) of $\approx 1.38 \times 10^{-3}$ Pa s. We can form a t-statistic (equation 3.1), $t = (1.45 - 1.38)/0.05 = 1.4$. The t is small; our estimated viscosity is close to the accepted value.

R.Box 3.8

A closer look at Reynolds' data

The diagnostic plots show that something happens at a pressure gradient of ≈ 65 Pa m^{-1}. We can repeat the regression using only the data at lower pressure gradients. To do this we either edit the data array(s) to remove these values and then repeat the regression, or we can use the `subset` argument to the `lm()` function as follows:

```
mask <- 1:8
result <- lm(v ~ dP, data = fluid, subset=mask)
plot(fluid$dP, fluid$v, bty="n", cex=1.5, pch=16)
abline(result)
plot(fluid$dP[mask], result$residuals, pch=16,
   xlab="Pressure grad (Pa/m)", ylab="residual (m/s)")
```

Now a plot of the data and model, and the plot of the residuals, suggest a much better match between data and model. The gradient of the line (see e.g. `summary(result)`) is 3.47×10^{-3}, with standard error of 0.12×10^{-3}.

We can also estimate the Reynolds number at the transition point:

$$Re = \frac{\rho v R}{2\eta}. \tag{3.24}$$

Inserting values of $v = 0.23$ m s^{-2} for the transition velocity, $\eta = 1.45 \times 10^{-3}$ Pa s as the dynamic viscosity, $R = 6.35 \times 10^{-3}$ m as the pipe radius and $\rho = 10^3$ kg m^{-3} as the density of water we get $Re = 2011$. Modern fluid dynamics, built on Reynolds' work, predicts that the transition to turbulence begins for $Re \gtrsim 2000$.

From the Reynolds' data (Appendix B, section B.3) and some simple linear regression analysis we have been able make several inferences. We can identify that at low velocity and pressure gradient a linear model (Poiseuille's equation) matches the data well, and estimate the values of velocity and pressure at which the laminar flow model breaks down. We have also been able to estimate the dynamic viscosity of water, and the Reynolds number at which the transition to turbulent flow occurs.

3.6.1 Going further with Reynolds' data

How could we learn even more from an experiment like this? Reynolds' data are not supplied with uncertainty estimates. In principle these could be estimated either by careful scrutiny of the experimental setup and analysis of all the likely sources of random fluctuations or uncertainty, or by repeating each estimate several times and using the sample of data points to estimate a mean measurement with standard error. Using equation 3.22 we can estimate the error to be $S = 6.30 \times 10^{-3}$ m s^{-1} for Reynolds' data. If we found the residual variance (SSE) was much larger than could be explained by the experimental errors, then we would need to think carefully about whether the linear model is suitable.

The simple linear regression method works well when the errors (the random component on each observation of the response variable) all have the same standard deviation.[5] We can perform a crude check of this by examining the residuals from the simple linear regression and checking for clear outliers. In this case the residuals do look reasonably symmetrical about the model, with no obvious outliers.

You may have also noticed that the simple linear model (equation 3.6) allows for a non-zero intercept, while Poiseuille's law, in the form of equation B.3, explicitly predicts a zero intercept. If the intercept were found to be significantly different from zero (e.g. using a t-test), then we would have an inconsistency between the prediction and the experimental results. Such an inconsistency could be due to a deficiency in the physical model, or some uncorrected bias introduced by the experiment.

R.Box 3.9

Numerical diagnostics of the Reynolds' data

We can compute the various sums of squares for a given regression analysis. In the case of the Reynolds data analysis, we can compute

```
v <- fluid$v[mask]
sst <- sum((v - mean(v))^2)
ssm <- sum((predict(result) - mean(v))^2)
sse <- sum((v - predict(result))^2)
r2 <- ssm/sst
cat(sst, ssm, sse, r2, fill=TRUE)
```

[5] In fact, they are assumed to be identically and independently normally distributed – these concepts will be dealt with later.

And we can check that indeed SST = SSM + SSE and that r^2 = SSM/SST gives the same r as the correlation coefficient (section 2.8):

```
(ssm + sse) / sst
sqrt(r2)
cor.test(fluid$dP[mask], fluid$v[mask])
```

This is a very high correlation coefficient, meaning the linear relationship accounts for the vast majority of the variance in the v data.

3.7 Chapter summary

- A one-sample t-statistic is used to quantify the disagreement between the mean of a sample and a predicted value.

$$t = \frac{\text{observed difference}}{\text{standard error}} = \frac{\bar{x} - \mu}{\sqrt{s_x^2/n}}.$$

- A two-sample t-statistic is used to quantify the disagreement between the means of two samples.

$$t = \frac{\text{difference of means}}{\text{standard error (of difference of means)}} = \frac{\bar{x} - \bar{y}}{\sqrt{s_x^2/n + s_y^2/n}}.$$

- Simple linear regression is a method to estimate the intercept and slope of a straight line that passes through the data points and minimises the sum of the squared (*data – model*) residuals. The intercept a and slope b estimates are

$$b = \frac{\overline{xy} - \bar{x}\bar{y}}{\overline{x^2} - \bar{x}^2} = r\frac{s_y}{s_x} \quad \text{and} \quad a = \bar{y} - b\bar{x}$$

  where r is the sample correlation coefficient (equation 2.6).
- After performing a regression analysis one should test the reasonableness of the fit by plotting the data and model, examining the residuals, and quantifying the amount of variance (in the response) variable explained by the model. In particular one can compute and compare three different variances:

$$\text{SST} = \sum_{i=1}^{n}(y_i - \bar{y})^2, \quad \text{SSM} = \sum_{i=1}^{n}(\hat{y}_i - \bar{y})^2, \quad \text{SSE} = \sum_{i=1}^{n}(y_i - \hat{y}_i)^2.$$

4

Probability theory

> They say that Understanding ought to work by the rules of right reason. These rules are, or ought to be, contained in Logic; but the actual science of Logic is conversant at present only with things either certain, impossible, or *entirely* doubtful, none of which (fortunately) we have to reason on. Therefore the true Logic for this world is the Calculus of Probabilities, which takes account of the magnitude of the probability (which is, or ought to be, in a reasonable man's mind).
>
> James Clerk Maxwell
>
> *(letter to L. Campell [June 1850])*[1]

What is the theory of probability and why is it useful? Probability theory shows us how to combine and manipulate probabilities for random experiments, and as such it underlies statistical analysis of random data.

4.1 Experiments, outcomes and events

Consider an experiment with a range of possible outcomes, but you are unable to predict which of the outcomes will occur. Classic examples include flipping a coin and drawing cards from a shuffled deck. The set that contains all the possible outcomes of the experiment is called the *sample space*, often given the label Ω, and each unique outcome is called an *element*. Some examples:

- If we flip a coin once there are two elementary outcomes, $\Omega = \{\mathrm{H}, \mathrm{T}\}$.
- If we flip a coin twice there are four elementary outcomes, $\Omega = \{\mathrm{HH}, \mathrm{HT}, \mathrm{TH}, \mathrm{TT}\}$.
- If we roll a six-sided die there are six elementary outcomes, $\Omega = \{$⚀, ⚁, ⚂, ⚃, ⚄, ⚅$\}$.

[1] See Campbell and Garnett (1882) for the rest of the letter, written when Maxwell was only 18.

We might not be so interested in elementary outcomes, but more general ones. For example, the outcome 'one head' from two flips of a coin is not elementary, as there are two ways this can happen, {HT, TH}. A set containing one or more outcomes is called an *event*, and we say that a particular event occurs if the outcome is in that set. We have little need to distinguish between individual outcomes (elementary outcomes) and sets of outcomes, so we shall use the term 'event' for both.

R.Box 4.1

Random sampling in R

R has the function `sample()` that simulates random sampling. We can create a sample space and randomly draw from it as many times as we like. For example, if we have 20 balls labelled 1, 2, ..., 20 we can draw five balls

```
S <- 1:20
sample(S, size=5)
```

The first line defines a sample space `S`, and the second line randomly draws a sample of size 5 from this space space. Note that, by default, the sampling is done *without* replacement; once element i has been drawn from the sample space it is effectively removed from the sample space and cannot be drawn again. We can use the `replace=TRUE` argument to draw with replacement (i.e. the sample space is returned to its original condition after each draw).

```
x <- sample(S, size=100, replace=TRUE)
```

R.Box 4.2

Frequency distribution of a random sample in R

Given some randomly sampled data `x` we can compute the frequencies of each event using the `hist()` command:

```
h <- hist(x, breaks=(0:20+0.5))
barplot(h$counts, space=1, names.arg=S)
```

The first line calculates (and by default plots) a histogram using bins centred on data value 1, 2, ..., 20. (We specify the break points between bins here.) But the data are discrete, not continuous, since we can obtain integer values only. We plot a barchart, with spaces between the bars, to emphasise the discreteness of the data. Try this again but make `x` a much larger sample (e.g. `size=1000` or `10000`) and see how the frequency distribution of the random sample changes as the sample size increases.

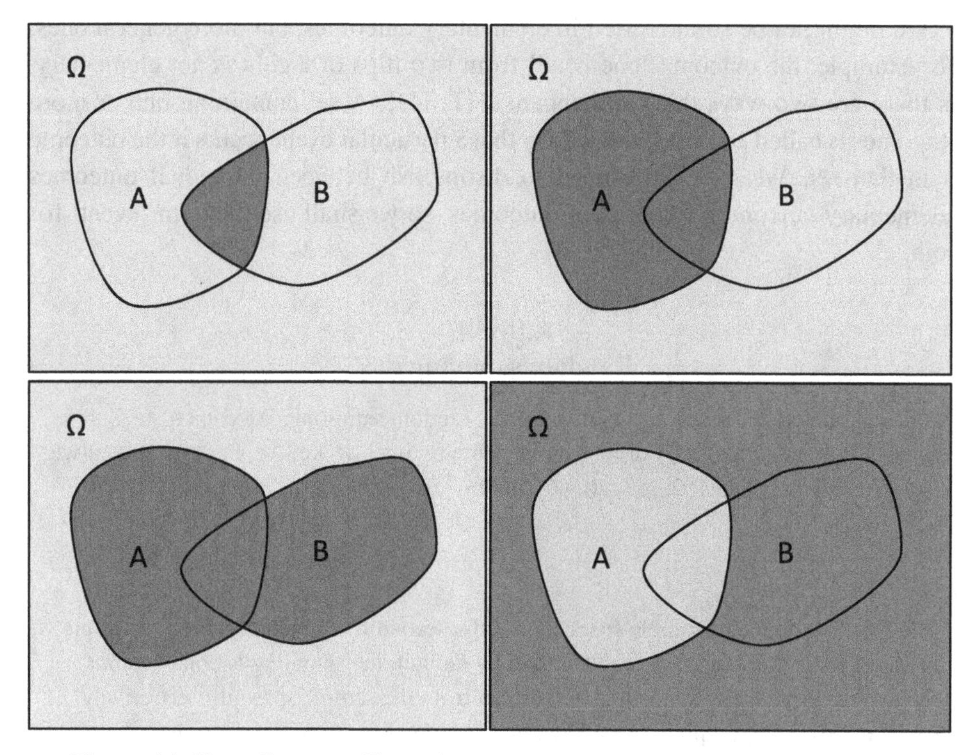

Figure 4.1 Venn diagrams illustrating events and their combinations. Ω is the sample space (the set of all possible outcomes), and A and B are events within this (each is a set of the possible outcomes). In the top-right panel, the event A is shaded; in the bottom-right panel, the event A^{C} (the complement of A) is shaded. In the top-left panel, the event $A \cap B$ (read 'A and B') is shaded; in the bottom-left panel, the event $A \cup B$ (read 'A or B') is shaded.

4.1.1 Combining events

We can collect several elementary outcomes into events, but there are more ways to combine outcomes and events. For these it is useful to use the language and notation of set theory. The basic combinations are illustrated by the Venn diagrams of Figure 4.1.

- The set $A \cap B$ is the *intersection* of the sets A and B (sometimes written just AB). It is the set of all elements that are in A and in B. The event $A \cap B$ is the event that occurs if both A and B occur; it occurs if the outcome is in A and in B. For example $\{1, 2, 3\} \cap \{2, 3, 4, 5\} = \{2, 3\}$ because 2 and 3 are the only elements common to both sets.
- The set $A \cup B$ is the *union* of the sets A and B. It is the set of all elements that are in either A or in B (or both). The event $A \cup B$ occurs if either A or

B occur, that is, if the outcome is in A or in B (or in both). For example $\{1, 2, 3\} \cup \{2, 3, 4, 5\} = \{1, 2, 3, 4, 5\}$.

- The set A^{C} is the *complement* of A, the subset of the sample space containing all the elements not in A. It corresponds to the event 'A does not occur'. In the example of the two coin flips, the complement of the event $B = \{\mathrm{HH, HT, TH}\}$ is $B^{\mathrm{C}} = \{\mathrm{TT}\}$.

The complement of an event is not necessarily its opposite. Consider the outcomes of rolling a die. If D is the event of rolling ⚃, then its complement D^{C} occurs if the die roll gives any other face, i.e. one of {⚀, ⚁, ⚂, ⚄, ⚅}.

It should be obvious that for any event A, either A or A^{C} (but not both) must occur. The outcome must either be in the set A or outside it; there are no other possibilities! In other words, the event 'A or A^{C}' is equal to the sample space: $A \cup A^{\mathrm{C}} = \Omega$ for any event A. Similarly, the event 'A and A^{C}' is impossible: $A \cap A^{\mathrm{C}} = \varnothing$ – there are no elements of the sample space that are in A and also in A^{C}.

Box 4.1

Summary of set notation for events

A = an event

B = another event

A^{C} = The compliment of A, i.e. the event 'not A'

Ω = The sample space of all possible events

$\varnothing$ = The null (empty) set, i.e. the set with no elements

$A \cap B$ = 'A and B' (intersection of sets)

$A \cup B$ = 'A or B' (union of sets)

Using this notation we may form new events by combining others. We could, for example label the outcomes of rolling a six-sided die A = ⚀ through to F = ⚅. We may then form a new event G from the union $G = E \cup F$, i.e. the event 'five or six' (which is the same as the event 'higher than four').

Two events, A and B, are said to be *mutually exclusive* if there are no outcomes that lie in both A and B (symbolically $A \cap B = \varnothing$). For example, when rolling a die the events A = ⚀ and F = ⚅ are mutually exclusive ($A \cap F = \varnothing$), but events F and G are not exclusive since both occur on the roll of ⚅.

R.Box 4.3
Playing cards in R

We can use the `sample()` function to simulate the drawing of playing cards. In a standard pack there are 52 different cards; each has a unique combination of rank ($A, 2, \ldots, 10, J, Q, K$) and suit (hearts, diamonds, clubs, spades).

We can define a *data frame* called `cards` that lists the 52 cards, produced by the possible the combinations of rank (which we shall label with numbers $1, \ldots, 13$) and suits. The `expand.grid()` function is used to populate the data frame with the combinations.

```
face <- 1:13
suit <- c("heart", "diamond", "spade", "club")
cards <- expand.grid(face, suit)
colnames(cards) <- c("face", "suit")
```

(Notice we have called the rank `face` to avoid confusion with a function `rank()`.) If you examine the contents of `cards` (just type its name on the command line) you should see the list of all 52 cards.

R.Box 4.4
Colour-matching cards in R

We shall find it useful to mark each card's colour (red or black). We can do this by testing the suit of the card, and depending on the result assigning it a colour

```
cards$colour <- (cards$suit == "heart" |
                 cards$suit == "diamond")
cards$colour[cards$colour == TRUE] <- "red"
cards$colour[cards$colour == FALSE] <- "black"
```

The symbol '|' means a logical 'or' and is used to combine two tests. The result of `x | y` is `TRUE` if either `x == TRUE` and/or `y == TRUE` are true. This produces another column called `colour`, which contains either `red` or `black` for each card.

R.Box 4.5
Picking playing cards in R

Now, we define the sample space as the integers $1, 2, \ldots, 52$, one for each card. We can then randomly draw these numbers, and find the corresponding cards as follows:

```
S <- 1:nrow(cards)
x <- sample(S, size=5)
cards[x, ]
```

This prints the appropriate rows (those listed in `x`) of the `cards` array, which contains the details of the cards. The space for the columns was left empty (`cards[x, ]`) to show all columns. If we wanted to print only the first two columns we could use e.g. `cards[x,1:2]`.

R.Box 4.6
Subsets in R

R has functions to compute, combine and compare subsets of a sample space. For example, the set of ace cards and the set of black cards can be found using:

```
setA <- S[cards$face == "1"]
setB <- S[cards$colour == "black"]
```

The result `setA` contains only the (row) numbers of the cards which satisfy the test criterion `card$face == "1"`. It is worth examining these to be sure you understand them

```
setA        # try also cards[setA, ]
setB        # try also cards[setB, ]
```

R.Box 4.7
Combining sets in R

We can combine sets, using specific set functions, to make other sets.

```
setAandB <- intersect(setA, setB)
setAorB <- union(setA, setB)
```

The result of the `intersect()` function is the set whose elements are common to both the input sets. The output of the `union()` function is the set whose elements are in either (or both) the input sets. It is worth examining the results:

```
setAandB    # try also cards[setAandB, ]
setAorB     # try also cards[setAorB, ]
```

4.2 Probability

The probability of A is a function $\Pr(A)$ whose value is a number in the range from 0 to 1. An impossible event has a probability of zero,[2] i.e. $\Pr(\varnothing) = 0$, and a certain

[2] Events that are impossible have zero probability, but events with zero probability are not always impossible. This one curious fact arises when we make the jump from discrete events to the continuous case. We will return to this in section 4.4.2.

event has a probability of unity, i.e. $\Pr(\Omega) = 1$. It should be fairly clear that since $A \cup A^{\mathrm{C}} = \Omega$ is certain then $\Pr(A \cup A^{\mathrm{C}}) = 1$, and likewise since $A \cap A^{\mathrm{C}} = \varnothing$ is impossible then $\Pr(A \cap A^{\mathrm{C}}) = 0$. However, most of the time we are concerned with situations in which the probability of an event takes a value between these two extremes, $0 \leq \Pr(A) \leq 1$. Towards the end of this chapter we shall consider in a little more detail the question of what probability actually *is*.

One of the basic rules of probability is that when two events, A and B, are mutually exclusive ($A \cap B = \varnothing$) the probability that either A or B occurs is the sum of their individual probabilities.

$$\Pr(A \cup B) = \Pr(A) + \Pr(B). \tag{4.1}$$

One, and only one, of the events A and A^{C} must occur, which means

$$\Pr(A \cup A^{\mathrm{C}}) = \Pr(A) + \Pr(A^{\mathrm{C}}) = 1 \tag{4.2}$$

and so

$$\Pr(A) = 1 - \Pr(A^{\mathrm{C}}). \tag{4.3}$$

This is a simple rule for manipulating probabilities, sometimes called the *complement rule*. Using this we can see that the probability of a die roll giving a number different from two is equal to Pr(⚁^{C}) = 1 − Pr(⚁).

Box 4.2
Odds

One often hears probability assessments in terms of the odds for event A, which is the ratio of the probabilities of A occurring, and A not occurring (i.e. A^{C} occurring):

$$odds(A) = \frac{\Pr(A)}{\Pr(A^{\mathrm{C}})} = \frac{\Pr(A)}{1 - \Pr(A)} \qquad \Pr(A) = \frac{odds(A)}{1 + odds(A)}$$

For example, odds of 15 : 1 against, or $odds = 1/15$, means the probability of success is $odds/(1 + odds) = 1/16 = 0.0625$. We leave it as an exercise for the reader to show that the latter equation follows from the former.

4.2.1 Conditional probability

Conditional probability is the probability of an event, given that some other event occurs. The conditional symbol is a vertical bar: $\Pr(A|B)$ means the probability of event A conditional on event B (or 'probability of A given B'). The definition of

Figure 4.2 Understanding conditional probability. (Credit: xkcd.com)

conditional probability is

$$\Pr(A|B) = \frac{\Pr(A \cap B)}{\Pr(B)} \tag{4.4}$$

(assuming that $\Pr(B) \neq 0$). It is worth investing some time getting to grips with the concept of conditional probability, arguably one of the most subtle but important ingredients of the probability calculus. If you understand it well much of what follows should be obvious.

The Venn diagram of Figure 4.1 (top-left) may help to make the meaning of $\Pr(A|B)$ a little more clear. If we know that event B does occur, we can forget about the rest of the sample space (B^{C}). And if event $A|B$ occurs then both A and B occur, i.e. the combined event $A \cap B$. So if we want to know the probability of $A|B$, we look at the probability of $A \cap B$ occurring from the reduced sample space B.

We can write down some simple probability relations:

$$\Pr(A|A) = 1$$
$$\Pr(A|A^{\mathrm{C}}) = 0. \tag{4.5}$$

These simply assert 'A is certain given A', and 'A is impossible given A^{C}', respectively. It should be noted that generally

$$\Pr(A|B) \neq \Pr(B|A). \tag{4.6}$$

These two are confused often enough that the mistake has been given a name: the *conditional probability fallacy*. One particularly important instance of this is the so-called prosecutor's fallacy. On the left side of the conditioning bar is the event we want the *probability of*, on the right side of the conditioning bar is the event we are *conditioning on*.

There are good reasons for considering all probabilities to be conditional, but often we neglect to note the conditioning explicitly. For example, we might consider 'the probability of rolling a six from a die' but that probability is conditional on the number of sides of the die and how it is rolled. If the die in question was 10 sided or 6 sided our sample space would be different, and so in general the probabilities would be different. Any probability is really conditional on the sample space that defines the possible outcomes, but often the sample space is assumed to be obvious from the context. However, failure to be explicit about the sample space is the cause of many problems and apparent paradoxes in statistics and probability.

Box 4.3
Example: Conditional probability

Imagine conducting a survey of pregnancy rates in the population – the sample space is therefore the set of all people in the area, which includes males and females. Let us say event G stands for 'person is pregnant' and event F stands for 'person is female'. It is then obvious that the probability of a person being pregnant given they are female, $\Pr(G|F) \approx 0.03$, is rather different from the probability of their being female given they are pregnant $\Pr(F|G) = 1$. The latter is a certainty but the former is not.

4.3 The rules of the probability calculus

We shall consider three rules of probability theory. We shall state these here and then consider them in more detail. All of probability theory can be derived from

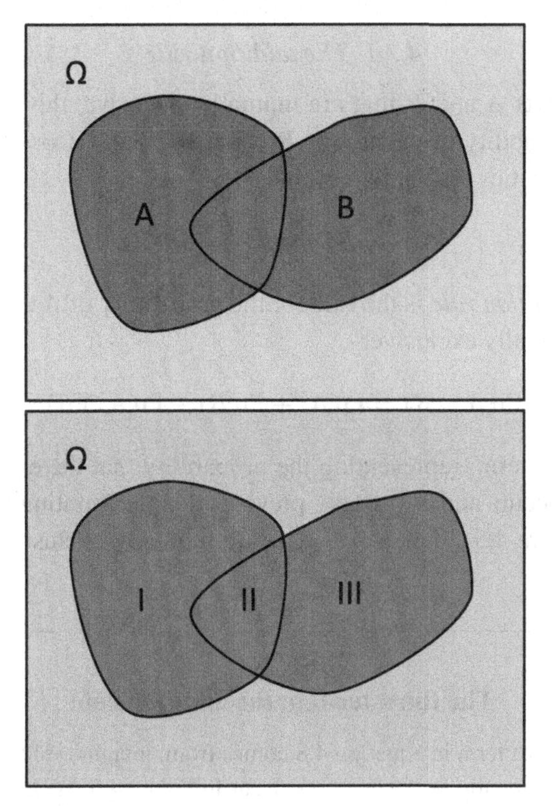

Figure 4.3 Venn diagram illustrating how the event $A \cup B$ (top) is divided into three sections, I, II and III (bottom).

these rules.[3]

Convexity rule:	$0 \leq \Pr(A\|B) \leq 1$ and $\Pr(A\|A) = 1$
Addition rule:	$\Pr(A \cup B) = \Pr(A) + \Pr(B) - \Pr(A \cap B)$
Multiplication rule:	$\Pr(A \cap B) = \Pr(A\|B)\Pr(B)$

where A and B represent discrete outcomes on a sample space Ω. The convexity rule is the convention that probability spans the range 0 to 1, with impossible events having probability $\Pr(\varnothing) = 0$ and certain events having probability $\Pr(\Omega) = 1$. We shall now discuss the other rules, and their implications, in more detail.

[3] In fact, the whole of probability theory, including the addition and multiplication rules, can be derived from three simple axioms due to Kolmogorov.

4.3.1 The addition rule

Consider two events A and B that are mutually exclusive, this is $A \cap B = \emptyset$. In this case, the probability of event $A \cup B$ occurring (i.e. 'A or B'), is simply the sum of the probabilities for the two events.

$$\Pr(A \cup B) = \Pr(A) + \Pr(B). \tag{4.7}$$

The general *addition rule* is the rule for the probability of $A \cup B$ even if the two events are not mutually exclusive:

$$\Pr(A \cup B) = \Pr(A) + \Pr(B) - \Pr(A \cap B). \tag{4.8}$$

Intuitively, the last term, representing the probability that the event $A \cap B$ occurs (i.e. that both A occurs and B occurs), prevents double counting when both A and B occur (see Figure 4.3). When A and B are mutually exclusive the last term is zero.

Box 4.4
The third term in the addition rule

To see where the last term in equation 4.8 comes from, imagine splitting the event $A \cup B$ into three mutually exclusive sections, as in Figure 4.3. We can see that $A = \mathrm{I} \cup \mathrm{II}$, $B = \mathrm{II} \cup \mathrm{III}$, $A \cap B = \mathrm{II}$ and $A \cup B = \mathrm{I} \cup \mathrm{II} \cup \mathrm{III}$. Since I, II and III are mutually exclusive it follows that

$$\begin{aligned}
\Pr(A \cup B) &= \Pr(\mathrm{I} \cup \mathrm{II} \cup \mathrm{III}) = \Pr(\mathrm{I}) + \Pr(\mathrm{II}) + \Pr(\mathrm{III}) \\
\Pr(A) &= \Pr(\mathrm{I} \cup \mathrm{II}) = \Pr(\mathrm{I}) + \Pr(\mathrm{II}) \\
\Pr(B) &= \Pr(\mathrm{II} \cup \mathrm{III}) = \Pr(\mathrm{II}) + \Pr(\mathrm{III}) \\
\Rightarrow \Pr(A \cup B) &= \Pr(A) + \Pr(B) - \Pr(A \cap B).
\end{aligned}$$

The general expression valid for any number of overlapping events, e.g. $\Pr(A \cup B \cup C)$, is given by the *inclusion–exclusion principle*.

Box 4.5
The probability of a black or an ace?

What is the probability that a card picked at random from a standard 52 card deck is either a black card or an ace? Our sample space is the set of all 52 cards: $\Omega = \{A_i\}$ with $i = 1, \ldots, 52$. Let us assign equal probability to each of the individual cards that could be picked, i.e. probability of drawing any particular card, A_i, is $\Pr(A_i) = q$ for all i. The addition rule for mutually exclusive events (equation 4.10) means

$$\Pr(\Omega) = \sum_{i=1}^{n} \Pr(A_i) = \sum_{i=1}^{n} q = nq = 1.$$

With $n = 52$ this gives the (expected) solution $q = 1/52$. We can calculate $\Pr(black \cup ace)$ using the addition rule,

$$\Pr(black \cup ace) = \Pr(black) + \Pr(ace) - \Pr(black \cap ace).$$

We can calculate each term individually by enumerating the possibilities as follows. $\Pr(black) = 26q$ since there are 26 black cards. $\Pr(ace) = 4q$ since there are four aces. $\Pr(black \cap ace) = 2q$ since there are only two cards that are both black and ace (A♠, A♣). The probability of drawing an ace or a black is therefore

$$\Pr(black \cup ace) = 26/52 + 4/52 - 2/52 = 28/52.$$

Notice how the third term in the sum prevents us from double-counting A♠ and A♣, which are included in the list of black cards and the list of ace cards. We can check this by counting how many of the 52 cards are either black or ace (or both), and indeed the number is 28.

Instead of just two events we may consider a larger set of n mutually exclusive events A_i $(i = 1, 2, \ldots, n)$ and get

$$\Pr(A_1 \cup A_2 \cup \cdots \cup A_n) = \sum_{i=1}^{n} \Pr(A_i). \tag{4.9}$$

If the events A_i are exhaustive as well as exclusive events (meaning one, and only one, must occur), then $A_1 \cup A_2 \cup \cdots \cup A_n = \Omega$, and so

$$\Pr(\cup_{i=1}^{n} A_i) = \sum_{i=1}^{n} \Pr(A_i) = \Pr(\Omega) = 1. \tag{4.10}$$

The total probability of exclusive and exhaustive events sums to unity (the union of exclusive and exhaustive events covers the whole sample space).

R.Box 4.8

Testing the addition rule in R

We can test equation 4.8 using the subsets defined above, using $A = ace$ and $B = black$. The probabilities for a card being in the set A, the set B, or the set $A \cap B$, are given by the numbers of elements in these sets compared to the sample space:

```
p.B <- length(setB)/length(S)
p.A <- length(setA)/length(S)
p.AandB <- length(setAandB)/length(S)
```

and the addition rule says the probability of drawing card that is ace or black is

```
p.A + p.B - p.AandB
```

This should match the probability calculated using the set of cards formed by the union of the set of aces and the set of blacks (R.box 4.7).

```
length(setAorB)/length(S)
```

4.3.2 The multiplication rule

The definition of the conditional probability (equation 4.4) may be rearranged to the *multiplication rule*

$$\Pr(A \cap B) = \Pr(A|B)\Pr(B). \tag{4.11}$$

In words this says the probability of 'A and B' is equal to the probability of 'A given B', multiplied by the probability of B. The individual events A, B are said to be *independent* if

$$\begin{aligned} \Pr(A|B) &= \Pr(A) \\ \Rightarrow \Pr(A \cap B) &= \Pr(A)\Pr(B). \end{aligned} \tag{4.12}$$

Equation 4.12 is a necessary and sufficient condition for A and B to be independent. (The correlation coefficient, discussed in sections 2.8 and 5.1.3, vanishes if A and B are independent, but a lack of correlation does not imply independence. Correlation is a weaker property than independence.)

The addition and multiplication rules (equations 4.8 and 4.11) are the two fundamental rules needed for manipulating probabilities; all other rules can be derived from these. We shall now consider three extensions of the basic rules, called 'extension of the conversation', 'law of total probability' and 'Bayes' theorem'.

Box 4.6
The probability of a black ace?

What is the probability that a card picked at random from a standard 52 card deck is a black ace? Given that $\Pr(A_i) = q = 1/52$ for $i = 1, \ldots, 52$, we simply need to count how many cards of the deck are both black and ace. The answer is two; they are A♠, A♣. Therefore $\Pr(black \cap ace) = 2q = 2/52$.

We could also calculate $\Pr(black \cap ace)$ using equation 4.11

$$\begin{aligned}\Pr(ace \cap black) &= \Pr(ace|black)\Pr(black)\\ &= (2/26)(1/2) = 2/52,\end{aligned}$$

where $\Pr(ace|black)$ is the probability of obtaining an ace drawing from only black cards, and $\Pr(black)$ is the probability of drawing a black card (from all 52 cards, i.e. Ω). Equivalently,

$$\begin{aligned}\Pr(black \cap ace) &= \Pr(black|ace)\Pr(ace)\\ &= (2/4)(4/52) = 2/52.\end{aligned}$$

R.Box 4.9

Testing the multiplication rule in R

We can test equation 4.11 using the subsets defined above, with $A = ace$ and $B = black$. We can calculate $\Pr(A|B)$ as follows:

```
setAgivenB <- setB[setB %in% setA]
cards[setAgivenB,]
p.AgivenB <- length(setAgivenB)/length(setB)
```

where `setAgivenB` is the subset of B whose elements are also in A. The conditional probability $\Pr(A|B)$ is then the number of elements in `setAgivenB` relative to the number in `setB`. Given this, the multiplication rule says that the probability of drawing a black ace is

```
p.AgivenB * p.B
```

which should be the same as the probability calculated using the set of cards formed by the intersection of the set of aces and the set of blacks.

```
length(setAandB)/length(S)
```

4.3.3 Extension of the conversation

What can we say about $\Pr(B)$ given that some other event A may have occurred? The event B occurs if either event $B \cap A$ or event $B \cap A^{C}$ occur: $B = (B \cap A) \cup (B \cap A^{C})$. Note that $B \cap A$ and $B \cap A^{C}$ are mutually exclusive, so their probabilities add

$$\begin{aligned}\Pr(B) &= \Pr((B \cap A) \cup (B \cap A^{\mathrm{C}})) \\ &= \Pr(B \cap A) + \Pr(B \cap A^{\mathrm{C}}) \\ &= \Pr(B|A)\Pr(A) + \Pr(B|A^{\mathrm{C}})\Pr(A^{\mathrm{C}}). \qquad (4.13)\end{aligned}$$

The last equality makes use of the multiplication rule (equation 4.11) to form an expression in terms of conditional probabilities $\Pr(B|A)$ and $\Pr(B|A^{\mathrm{C}})$. This way of expanding B into '$B \cap A$ or $B \cap A^{\mathrm{C}}$' is the rule known as 'extension of the conversation' and shows us how we can relate probabilities involving $\Pr(B|A)$ to $\Pr(B)$.

4.3.4 Law of total probability

The *law of total probability* (sometimes known as the total probability theorem) is similar to the extension of the conversation (equation 4.13). But instead of considering B conditional on two events (A and A^{C}), we can now include many exclusive and exhaustive events. We partition the sample space into $\Omega = \{A_1, A_2, \ldots, A_n\}$ and then construct the sum of the probabilities of B conditional on each of these,

$$\begin{aligned}\Pr(B) &= \Pr(B \cap \Omega) \\ &= \Pr((B \cap A_1) \cup (B \cap A_2) \cup \cdots \cup (B \cap A_n)) \\ &= \Pr(B \cap A_1) + \Pr(B \cap A_2) + \cdots + \Pr(B \cap A_n) \\ &= \sum_{i=1}^{n} \Pr(B \cap A_i) \\ &= \sum_{i=1}^{n} \Pr(B|A_i)\Pr(A_i). \qquad (4.14)\end{aligned}$$

This process of eliminating the conditionals (by summing over all possible values of the conditional variable) is called *marginalisation* for historical reasons (the summation was usually performed in the margins when written down on paper).

4.3.5 Bayes' theorem

The relation between $\Pr(A|B)$ and $\Pr(B|A)$ is given by *Bayes' theorem*, which is a simple consequence of the multiplication rule (equation 4.11). First, we note that the $A \cap B$ is the same as $B \cap A$, and then write the multiplication law in terms of

$\Pr(A \cap B)$ or $\Pr(B \cap A)$:

$$\Pr(A \cap B) = \Pr(B \cap A) = \Pr(A|B)\Pr(B) = \Pr(B|A)\Pr(A). \tag{4.15}$$

The last part can be rearranged to arrive at Bayes' theorem[4] (assuming that $\Pr(B) \neq 0$).

$$\Pr(A|B) = \frac{\Pr(B|A)\Pr(A)}{\Pr(B)}. \tag{4.16}$$

We can extend this rule to cover more than two events. Let us denote a set of exclusive and exhaustive events $\{A_1, A_2, \ldots, A_n\}$, then we can write Bayes' theorem as

$$\Pr(A_i|B) = \frac{\Pr(B|A_i)\Pr(A_i)}{\Pr(B)} = \frac{\Pr(B|A_i)\Pr(A_i)}{\sum_{i=1}^{n}\Pr(B|A_i)\Pr(A_i)}. \tag{4.17}$$

This means that if we know the probability assignments $\Pr(B|A_i)$ and $\Pr(A_i)$ we may calculate $\Pr(A_i|B)$.

4.3.6 A medical example

The following is an example of how Bayes' theorem may be used to calculate probabilities. (This is a popular example for illustrating Bayes' theorem; similar examples appear in some of the books listed in the references.)

Imagine testing for a disease (D) which is carried by 1% of the population; in other words we have

$$\Pr(D) = 0.01, \qquad \Pr(D^{\mathrm{C}}) = 0.99.$$

There is a very effective test for the disease. For a person that carries the disease, the probability of the test producing a positive result (a true positive) is 0.99

$$\Pr(+|D) = 0.99, \qquad \Pr(-|D) = 0.01.$$

On the other hand, for a person that does not carry the disease there is a 0.02 probability of producing a positive result (a false positive).

$$\Pr(+|D^{\mathrm{C}}) = 0.02, \qquad \Pr(-|D^{\mathrm{C}}) = 0.98.$$

The question is: if you are randomly picked to receive a test, and it returns a positive result, what is the probability you carry the disease? Let's apply Bayes' theorem to

[4] This is attributed to Rev. Thomas Bayes (1702–1761), whose posthumous publication *An Essay towards Solving a Problem in the Doctrine of Chances* (1763) included a version of the theorem. Laplace later (1812) rediscovered and generalised the theorem.

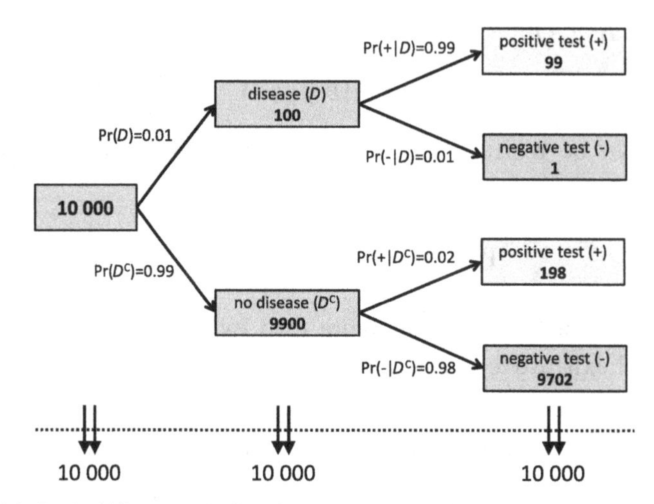

Figure 4.4 Probability tree indicating the combinations of events in the medical example. Reading from left to right, each branch of the tree indicates a possible event, conditional on any events preceding it. The probability of a compound event on the right (e.g. 'positive test' and 'disease') is the product of the probabilities along each branch taken to reach that event. The numbers in each box show the expected numbers if we start with a population of 10 000 people. At each branching, the size of the population (i.e. the sum over all possibilities) remains the same. In total there are $99 + 198 = 297$ positives from a population of 10 000, only 99 of which are true positives.

calculate the probability of having the disease given a positive test result:

$$\Pr(D|+) = \frac{\Pr(+|D)\Pr(D)}{\Pr(+)} = \frac{0.99 \times 0.01}{0.0297} = \frac{1}{3}$$

where

$$\begin{aligned}\Pr(+) &= \Pr(+|D)\Pr(D) + \Pr(+|D^C)\Pr(D^C) \\ &= 0.99 \times 0.01 + 0.02 \times 0.99 = 0.0297.\end{aligned}$$

This means the probability that you carry the disease, given you receive a positive test result, is 33.3%. This might seem surprising since the test is rather accurate (the probability of receiving a wrong result is only 1% or 2%, depending on whether you do or do not carry the disease). In particular, note that $\Pr(+|D) \neq \Pr(D|+)$. This type of mistake is known as 'the base rate fallacy' because it stems from not properly taking into account the base rate of the event we are interested in. In this case the low base rate of 1% incidence of the disease means the number of false positives will be at least as large as the number of true positives.

Still not convinced? Let's see what happens when we put things into real numbers. Imagine a population of $n = 10\,000$ people, all of whom take the test. The incidence of the disease is such that we expect $n \times \Pr(D) = 100$ members of the population to carry it, and so the remaining 9900 do not. Of the 100 people that carry the disease, we expect the number of positive tests to be $100 \times \Pr(+|D) = 99$ and the number of negative tests to be 1. In contrast, of the 9900 people who do not carry the disease we expect $9900 \times \Pr(+|D^{C}) = 9702$ negative tests, and by similar reasoning 198 positive tests. Now, all we know is that you have received a positive test, and so are one of the $99 + 198 = 297$ people who gave a positive test, only 99 of whom do in fact carry the disease. The probability that you have the disease, given that you are one of this population, is $\Pr(D|+) = 99/(99 + 198) = 1/3$ as before. The various events and their probabilities can be illustrated using a probability tree as in Figure 4.4.

4.3.7 A particle physics example

So the above medical example is not enough like physics for you, heh? We can see how the same logic applies if we consider particle detection. We have a particle detector that is designed to detect muons (μ particles), which it does with probability 0.99, but will also trigger on pions (π particles), with probability 0.02. If we use '+'='detector trigger' and '−'='no detector trigger', then we can write

$$\Pr(+|\mu) = 0.99 \;\; [\text{therefore } \Pr(-|\mu) = 0.01]$$

$$\Pr(+|\pi) = 0.02 \;\; [\text{therefore } \Pr(-|\pi) = 0.98].$$

Now if we place the particle detector in a beam comprising 99% π and 1% μ particles – so that $\Pr(\pi) = 1 - \Pr(\mu) = 0.99$ – what is the chance that a detection was caused by a μ? The problem is exactly the same as in the example of the medical test:

$$\Pr(\mu|+) = \frac{\Pr(+|\mu)\Pr(\mu)}{\Pr(+)} = \frac{\Pr(+|\mu)\Pr(\mu)}{\Pr(+|\mu)\Pr(\mu) + \Pr(+|\pi)\Pr(\pi)} = \frac{1}{3}.$$

The probability that a detection is really due to a μ particle is 1/3. Despite the much higher detection efficiency for μ particles, the majority of detector triggers are caused by π particles, due to the much higher probability of a π hitting the detector (the base rate fallacy).

Now, let us consider what happens when we demand that for a particle to be identified it must trigger two independent detectors. (We use ++ to indicate the event 'two triggers', i.e. 'detector A triggers and detector B triggers'.) Because the two detectors are independent, we may use the multiplication rule for independent events (equation 4.12) to give the probability of two triggers. Given a μ we have the probability of two triggers $\Pr(++|\mu) = \Pr(+|\mu)\Pr(+|\mu) = 0.98$. But given a π

we have the probability of two triggers $\Pr(++|\pi) = \Pr(+|\pi)\Pr(+|\pi) = 0.0004$. If we record a double trigger, the probability it is caused by a μ is then

$$\begin{aligned}\Pr(\mu|++) &= \frac{\Pr(++|\mu)\Pr(\mu)}{\Pr(++)} \\ &= \frac{\Pr(++|\mu)\Pr(\mu)}{\Pr(++|\mu)\Pr(\mu) + \Pr(++|\pi)\Pr(\pi)} \\ &= 0.96.\end{aligned}$$

This is a massive improvement in the quality of particle identification: a double trigger much more likely due to a μ than a π. One can imagine adding yet more detectors to improve the identification further. What are the drawbacks of such an approach? (Consider how $\Pr(\mathit{detection}|\mu)$ changes as more triggers are required for a detection, and what this means for the overall detection efficiency.)

4.4 Random variables

Technically, a *random variable* is a function that maps the sample space Ω of some random process onto real numbers. (A random process is one for which the outcome cannot be predicted in advance.) For example, the sample space corresponding to the flipping of a coin might comprise just two elementary events: $\Omega = \{\mathit{heads}, \mathit{tails}\}$. We can define a variable X

$$X = \begin{cases} 0 & \text{if tails} \\ 1 & \text{if heads.} \end{cases} \tag{4.18}$$

Its value depends on which random event occurs. Similarly, for a six-sided die we can consider the sample space comprising six elementary events $\Omega = \{$⚀, ⚁, ⚂, ⚃, ⚄, ⚅$\}$ and a random variable Y that takes on the value corresponding to each of these events, i.e.,

$$Y = \begin{cases} 1 & \text{if ⚀ is rolled} \\ 2 & \text{if ⚁ is rolled} \\ 3 & \text{if ⚂ is rolled} \\ 4 & \text{if ⚃ is rolled} \\ 5 & \text{if ⚄ is rolled} \\ 6 & \text{if ⚅ is rolled.} \end{cases} \tag{4.19}$$

The events in the sample space and the value of a random variable are different things. In our coin example the event *heads* is not the same as the variable X

taking the value 1, which is just a number. In mathematical jargon, there is a *mapping* between the events and the real numbers, and this mapping is called a random variable (mathematicians write this symbolically as $X : \Omega \to \mathbb{R}$). In a large fraction of what follows we shall be considering random variables.

4.4.1 Discrete random variables

At the risk of labouring the point we shall describe the notation in more detail. If you prefer to skip the technicalities you may safely pass over this subsection.

If ω is an elementary event from the sample space Ω we say that $\omega \in \Omega$ (read 'ω is a member of Ω') and the probability of the event occurring is $\Pr(\omega)$. Our random variable maps events to real numbers: $X(\cdot)$ is a function whose input argument is an event and whose output is a real number. The elements of Ω for which X takes on a particular value x can be written $\{\omega; X(\omega) = x\}$. We can therefore write down the probability that the variable X is equal to some particular number x as

$$\Pr(\{\omega; X(\omega) = x\}). \tag{4.20}$$

This is 'the probability of event ω, where ω is chosen such that $X(\omega) = x$'. But usually we are not interested in the nature of the events themselves (i.e. the particular ω), only the value of the variable. The events themselves are usually ignored from the probability notation and instead we may write the probability that the variable X takes the value x as

$$\Pr(X = x). \tag{4.21}$$

Following convention we use upper case (X) for the random variable and lower case (x) for values it takes. To simplify the notation further we can omit mention of the random variable itself and simply write

$$p(x) = \Pr(X = x) = \Pr(\{\omega; X(\omega) = x\}), \tag{4.22}$$

where a lower case $p(\cdot)$ is used to denote the probability function for some random variable. We have saved ourselves some considerable clutter by dropping mention of events such as ω and variables such as X. But this shorthand may present problems when we are discussing more than one random variable, say, X and Y. It may not be clear whether $p(2)$ refers to $X = 2$ or $Y = 2$. If the random variable is not clear from the context, we shall use a subscript to clarify

$$p_X(x) = \Pr(X = x). \tag{4.23}$$

Read: the probability that X takes on the value x. Often we need hardly mention the random variables. Instead it is quite straightforward to write everything using terms like $p(x)$ and $p(y)$, and insert numerical values when necessary. We shall use this notation throughout most of the rest of the book.[5] When we have need to be more definite about the probability of a random variable we shall use $\Pr(X)$.

Now, we can combine probabilities for discrete random variables just as for events. For example, the probability that $X \leq x$ is

$$F(x) = \Pr(X \leq x) = \sum_{x_i \leq x} \Pr(x_i) \tag{4.24}$$

where we sum the probabilities over all the values that X may take that are less than or equal to x. This gives the *cumulative distribution function* (cdf) for the variable.

4.4.2 Continuous random variables

In the physical sciences we very often consider the properties we measure to be continuous, and represented as discrete only because of the finite precision of the measuring and recording processes. For example, if we measure the sky position of a star, the wind speed at some time and location, or the lifetime of a transistor, we expect these to be continuous variables.

Here's the rub: if we have a continuous random variable X, the probability that it equals some number x exactly is zero, i.e. $\Pr(X = x) = 0$, because for a continuous variable there are infinite possible values it can take. If a random variable can take on any real number in the range 0 to 1, the sample space contains an infinite number of events (one for each number). Although any event in the sample space is possible, the chance of picking any particular event from that sample space is zero. (This is why zero probability does not imply impossibility.) Imagine throwing a dart into a board. The probability that the dart will hit any particular point on the board will be proportional to the size of the dart's tip relative to the size of the board. If we make the tip smaller and smaller, the probability of the dart landing on any particular point vanishes, even though the dart lands somewhere on the board.

But all is not lost. We can still define some definite probabilities in the continuous case, such as the probability of finding X, a continuous random variable, smaller than some value x. Call this

$$F(x) = \Pr(X \leq x). \tag{4.25}$$

[5] The thing to note is that $p(x)$ means the probability function for some random variable, which may be a different function from $p(z)$. We use $p(\ldots)$ as shorthand for any probability function, and have to keep track of which function applies to which variables.

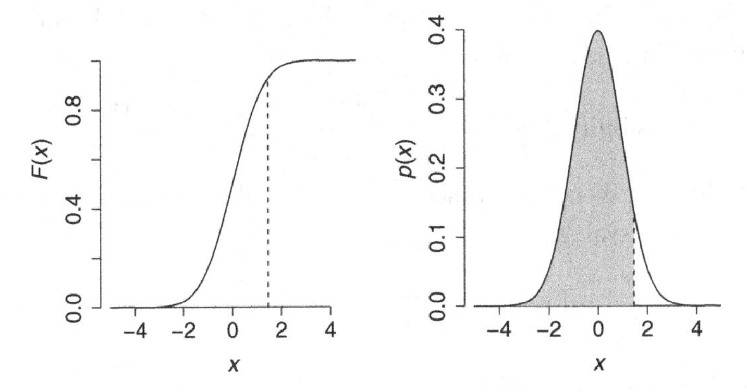

Figure 4.5 A cumulative distribution function (cdf; left) and the corresponding probability density function (pdf; right). The cumulative distribution corresponds to the running integral of the pdf (the area under the pdf curve from left to right).

See Figure 4.5. This is usually known as the *cumulative distribution function* (cdf). It gives the probability of '$X \leq x$' for some value x. As such it obeys the rules we would expect for a probability, notably $0 \leq F(x) \leq 1$. In the limits

$$\lim_{x \to -\infty} F(x) = 0 \qquad \lim_{x \to +\infty} F(x) = 1. \tag{4.26}$$

$F(x)$ is a monotonic, non-decreasing function of x such that $F(x_1) \geq F(x_0)$ for all $x_1 > x_0$. We can use this to find the probability that X lies in some particular range

$$\Pr(a \leq X \leq b) = F(b) - F(a). \tag{4.27}$$

If we make the interval[6] $[a, b]$ smaller then the probability decreases. The probability that '$X = x$' is zero, but we can consider the finite probability of X being in some small interval $[x, x + \delta x]$ and calculate the probability density

$$\frac{\Pr(x \leq X \leq x + \delta x)}{\delta x} = \frac{F(x + \delta x) - F(x)}{\delta x}, \tag{4.28}$$

and then take the limit, called the *probability density function* (pdf), or sometimes just density

$$p(x) = \lim_{\delta x \to 0} \frac{\Pr(x \leq X \leq x + \delta x)}{\delta x} = \frac{\mathrm{d}F(x)}{\mathrm{d}x}. \tag{4.29}$$

[6] Note that square brackets are used to denote an *open interval*, one that includes its endpoints. So $[a, b]$ means any value between a and b, including a and b. Curved brackets indicate a *closed interval*, one that does not include its endpoints. Thus (a, b) includes values between a and b, but excludes points a and b. For a continuous case we can be relaxed about the specific type of interval since the end points contribute zero probability.

Figure 4.5 illustrates the difference between the cumulative and density functions. We use 'density' in a manner analogous to its use in physics,

$$\text{probability density} = \frac{\text{probability 'mass' in interval}}{\text{interval size}}. \tag{4.30}$$

The probability that X occurs in some interval is then given by integrating the density over the interval, e.g.,

$$\Pr(X \leq x) = F(x) = \int_{-\infty}^{x} p(x')\mathrm{d}x' \tag{4.31}$$

(where x' is a dummy variable), and

$$\Pr(a \leq X \leq b) = F(b) - F(a) = \int_{a}^{b} p(x)\mathrm{d}x, \tag{4.32}$$

and the density is normalised in a manner analogous to the discrete case (see equation 4.10) such that

$$\int_{-\infty}^{+\infty} p(x)\mathrm{d}x = 1, \tag{4.33}$$

which simply says $\Pr(-\infty \leq X \leq +\infty) = 1$. A pdf must also be non-negative, $p(x) \geq 0$ wherever $p(x)$ is defined. It is possible for a probability density $p(x) > 1$, as long as it still integrates to unity (see equation 4.33). The set of values that a random variable may take with non-zero probability (i.e. where $p(x) > 0$) is called its *support*.

In section 2.5 we used quantiles to describe samples of data. The inverse of a cdf $F^{-1}(x)$ gives the quantiles of a distribution. The α-point quantile x_α is

$$F(x_\alpha) = \int_{-\infty}^{x_\alpha} p(x)\mathrm{d}x = \alpha \quad \Longleftrightarrow \quad x_\alpha = F^{-1}(\alpha). \tag{4.34}$$

The most well-known quantile is the median, the $\alpha = 0.5$ quantile or $x_{0.5}$, the value of the variable X for which $F(x) = 0.5$. But in principle we can work with any other quantile $0 < \alpha < 1$. For example, the interquartile range (IQR) of a distribution is $x_{0.75} - x_{0.25}$.

It turns out that the rules we have for probabilities are almost exactly the same for probability densities. For example, Bayes' theorem and the multiplication rule look the same. The main difference is that, where we sum over discrete outcomes in the discrete case, we must integrate over a continuous range of outcomes in the continuous case. For this reason most writers use the same symbol, usually $p(\cdot)$, for both probability (mass) and probability density, and where necessary state explicitly whether the variable of interest is discrete or continuous. We shall follow that practice here.

4.4.3 Two random variables

Pdfs may be functions of any number of continuous random variables, or may be mixed functions of continuous and discrete random variables. In the same way that we approached the pdf of a single continuous variable we can consider the *joint* probability density function of two variables, X and Y:

$$p_{X,Y}(x, y) = \lim_{\delta x, \delta y \to 0} \frac{\Pr(x \leq X \leq x + \delta x \text{ and } y \leq Y \leq y + \delta y)}{\delta x \delta y}. \tag{4.35}$$

Notice that we use (x, y) to denote dependence on both x and y (rather than the notation used above for joint events, e.g. $A \cap B$).

In general, the probability of random variables X and Y having values in some region R is

$$\Pr(X \textit{ and } Y \textit{ in } R) = \iint_R p_{X,Y}(x, y) \mathrm{d}x \mathrm{d}y. \tag{4.36}$$

(Compare with equation 4.32.) The integrals are replaced by sums for any variables that are discrete. There is also a continuous form of equation 4.14 that allows us to calculate a marginal density (or pdf).

$$p_X(x) = \int_{-\infty}^{+\infty} p_{X,Y}(x, y) \mathrm{d}y. \tag{4.37}$$

This is the projection of the surface density $p_{X,Y}(x, y)$ onto the x axis. It tells us the probability density for $X = x$, over all possible values of Y. We form the marginal density $p_Y(y)$ in an analogous fashion. Marginalization therefore eliminates one or more variables from the density function, leaving a density that is a function of a reduced number of variables.

We can also consider conditional probability densities such as $p_{X|Y}(x\,|\,y_0)$. This tells us the probability density for $X = x$ given that $Y = y_0$; it is a 'slice' through the joint density along the line $Y = y_0$.

$$p(x\,|\,y_0) = \frac{p(x, Y = y_0)}{p(y_0)} = \frac{p(x, Y = y_0)}{\int p(x, Y = y_0) \mathrm{d}x}. \tag{4.38}$$

This is analogous to the case for discrete variables (equation 4.4). The denominator is the marginal density of Y (see equation 4.37) evaluated at $Y = y_0$, and ensures that the above density integrates to unity. See Figure 4.6. Notice that in order to avoid clutter we have not explicitly noted which random variables are being considered, as this should be obvious from the context, e.g. $p(x) = p_X(x)$ and $p(x, y) = p_{X,Y}(x, y)$.

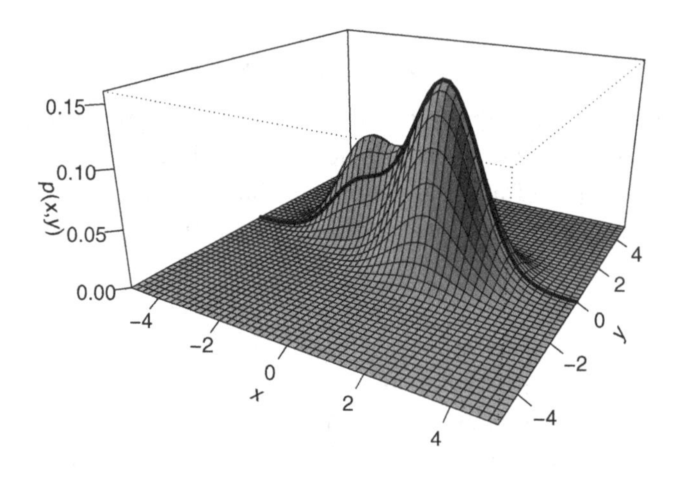

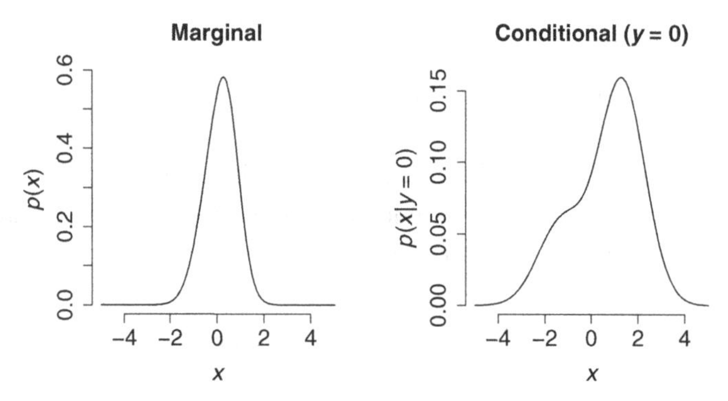

Figure 4.6 Example of a joint probability distribution function $p(x, y)$ of two random variables (top), with the function traced through $y = 0$ (thick line). Also shown are the marginal distribution $p(x) = \int p(x, y)\mathrm{d}y$ (bottom left) and conditional distribution $p(x|y = 0)$ (bottom right) corresponding to the trace.

The multiplication rule (equation 4.11) can be written in continuous form using the definition of the conditional probability (equation 4.38):

$$p(x, y) = p(y, x) = p(x|y)p(y) = p(y|x)p(x). \tag{4.39}$$

Two random variables X and Y are said to be *independent* if and only if

$$p(x, y) = p(x)p(y) \tag{4.40}$$

(compare with equation 4.12). We may therefore rewrite the marginal distribution of equation 4.37, using equation 4.39, as

$$p(x) = \int_{-\infty}^{+\infty} p(x|y)p(y)\mathrm{d}y \tag{4.41}$$

which is the continuous form of equation 4.14.

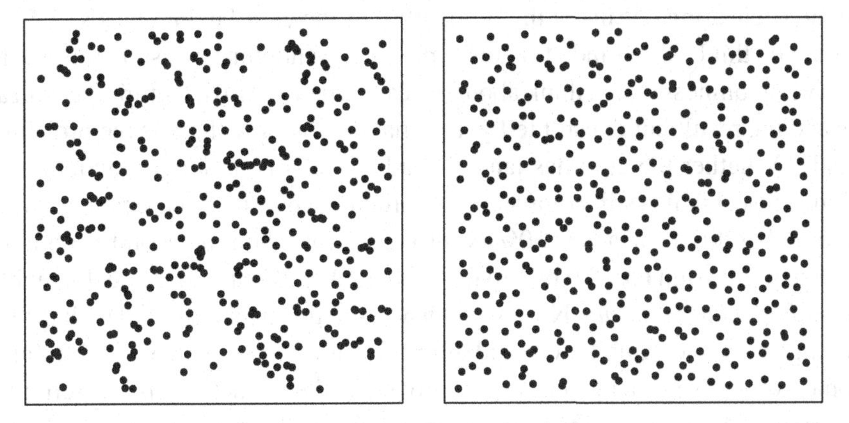

Figure 4.7 Two examples of random images.

Exactly as in the discrete case, where the multiplication rule was re-arranged to produce Bayes' theorem, the continuous multiplication rule (equation 4.39) leads to Bayes' theorem for continuous variables:

$$p(y|x) = \frac{p(x|y)p(y)}{p(x)} = \frac{p(x|y)p(y)}{\int_{-\infty}^{+\infty} p(x|y)p(y)\mathrm{d}y} \tag{4.42}$$

(compare with equation 4.17).

4.5 The visual perception of randomness

Figure 4.7 shows two images, each of approximately 400 randomly positioned dots. In one image the dots are randomly distributed uniformly along each axis, in the other image there is some additional (statistical) structure. But which is which? For the image on the left, the x and y positions of the dots are independent, uniformly distributed random numbers. The image on the right contains structure that artificially spreads the points more evenly than you would expect if they were randomly and uniformly distributed. The lesson: expect random clusters and voids even in completely random data.

4.6 The meaning of 'probability' and 'random'

Now we return to the more philosophical but nevertheless important issue of what 'probability' actually *is*, or what 'randomness' actually *means*.

First of all let's consider what we mean when we say something is random. Think about watching a friend about to flip a coin. Is the outcome (heads or tails) random? Surely it is random because it is practically impossible to predict (unless someone is cheating). What is the probability of the coin landing heads up? Fifty per cent, you think. This kind of (natural, physical) randomness is sometimes called *aleatory* randomness. Now, think about the case when your friend flips the coin and looks and the result but doesn't tell you. What do you now think is the probability of heads? Whether the coin has landed heads or tails is no longer random in the previous sense, but it is still completely uncertain to you. In this case you might still say the probability of heads is 50% (your friend, for whom the result is no longer uncertain, will assign either 0 or 1 to the probability). What you are dealing with is *epistemological uncertainty*. Now, your answers to the questions (and perhaps your reasoning) were the same before and after the coin flip, so what's the difference? To some extent it's just a matter of perspective whether something is indeterminate, or determined but unknown to you. The same probability theory can be applied in both cases.

In practice there are different approaches that limit where the concepts of 'probability' and 'randomness' are applied. There is the mathematical theory of probability, based on a system of axioms (rules) for manipulating sets or variables, but this doesn't say anything about how to apply the concept of probability to real problems. There are two main schools of thought on the application of probability theory and they differ fundamentally on what a *probability* actually is. There is the *frequentist* interpretation (sometimes referred to as classical or orthodox), and the *Bayesian* interpretation (sometimes referred to as modern or subjective). (Actually, there are differences of interpretation within these schools, but we shall ignore those here.) The difference is not just a technicality; it affects how probability theory may be applied in making inferences. Below we briefly outline the two definitions, but for a fuller discussion see, e.g., Berger and Berry (1988), Howson and Urback (1991) and Jeffreys and Berger (1992).

Frequency definition: this says that probability is the relative frequency of events in the 'long run'. Consider an experiment in which n events are observed, and some of these events are of type X. The *probability* of X is the relative frequency that event X occurs in an infinite sequence:

$$\Pr(X) = \lim_{n\to\infty} \frac{\text{number of times event } X \text{ occurs in } n \text{ experiments}}{n}. \tag{4.43}$$

Of course, one has to be willing to assume that the limit exists. Within this scheme is it valid to assign probabilities to things such as radioactive decays, lottery draws and coin flips. In all these cases one can imagine a long series of practically identical repeats, and the probability of an outcome is the limit of the frequency of that outcome.

By contrast, it is not meaningful to consider the probability that it will rain on a certain date, because that date will only happen once and cannot be repeated.

Likewise, one cannot assign a probability to hypotheses about the mass of Saturn, or whether an asteroid impact caused a mass extinction 65 million years ago. Saturn has a mass that is not intrinsically random but it is uncertain (to us); an asteroid impact either did or did not trigger the Cretaceous–Paleogene mass extinction. It is rather difficult to imagine a series of repeats for these from which to define a long-run frequency and hence probability. The frequentist approach sticks firmly to aleatory randomness.

Bayesian definition: The *Bayesian*[7] probability is interpreted more broadly as a model for uncertainty. It allows probabilities to be used to quantitatively handle (aleatory) randomness and (epistemological) uncertainty. As such, it is valid to consider the probability for a hypothesis, and interpret this as a measure of uncertainty. Our knowledge of the hypothesis is not random, but it is uncertain.

Within the Bayesian scheme, it is valid to consider the probability that it will rain on a certain date, the probability for the hypothesis that an asteroid impact was responsible for the Cretaceous–Paleogene extinction, and assign a probability function to the mass of Saturn. These probabilities do not reflect the relative frequencies of events occurring; instead, they reflect a quantitative assessment of our uncertainty about them, given the available evidence (data) and prior knowledge of the situation. In our daily and professional lives, most of us use probability in this sense; when reasoning from a state of uncertainty we often use probability to express our uncertain knowledge (e.g. 'it probably will rain this weekend'). The Bayesian approach includes both (aleatory) randomness and (epistemological) uncertainty and treats them both using the probability calculus.

4.6.1 Different strokes...

Frequentists do not assign probabilities to hypotheses or parameters, only to random variables (such as noisy data). For example, imagine searching for a new particle in noisy data: what is the probability of the particle's existence? Either there is a new particle or there is not; this is not strictly a random variable and so cannot be assigned a probability in the frequentist view. The frequentist must imagine repeating the experiment an infinite number of times, in order to represent the population of possible datasets from which the real data were drawn. The frequentist can discuss the probability of recording data like those actually taken, under different assumptions (e.g. the particle is or is not there). Frequentist analysis speaks in terms of the probability of random data given certain hypotheses; it does not allocate probabilities to the hypotheses given actual data. In contrast,

[7] For more on the origins and usage of the word 'Bayesian', see Fienberg (2006), available via www.stat.cmu.edu/~fienberg/fienberg-BA-06-Bayesian.pdf.

the Bayesian works towards the probability of hypotheses (e.g. there is a new particle).

In science we are often interested in making inferences about hypotheses. To do so we may either adopt the Bayesian approach and consider directly the probability of relevant propositions, or take the frequentist approach and consider only probabilities of random data, then use additional arguments to infer something about the hypotheses under scrutiny.

There are now many good books describing Bayesian approaches to data analysis. The books by Sivia and Skilling (2006) and Gregory (2005) are written by and for physicists. Bolstad (2007) gives a broader introduction, at a relatively simple level, while Gelman *et al.* (2003) and Lee (2004) give more detailed treatments. Albert (2007) covers computational aspects (with example R code).

4.7 Chapter summary

- *Frequentist probability* The long-run relative frequency of event X: $\Pr(X) = \lim_{n\to\infty} n_X/n$.
- *Bayesian probability* A quantitative measure of uncertainty about a hypothesis X.
- Combining events: 'not', 'and', 'or' and 'conditional'

$$A^{\mathrm{C}} = \text{not } A \qquad A \cap B = A \text{ and } B$$
$$A \cup B = A \text{ or } B \qquad A|B = A \text{ given } B.$$

- The three rules of the probability calculus

Convexity rule:	$0 \leq \Pr(A\|B)$ and $\Pr(A\|A) = 1$
Addition rule:	$\Pr(A \cup B) = \Pr(A) + \Pr(B) - \Pr(A \cap B)$
Multiplication rule:	$\Pr(A \cap B) = \Pr(A\|B)\Pr(B)$.

- Probability density functions (pdfs) for a continuous random variable X and the joint density function for the pair of variables X, Y i.e. the probability densities at $X = x$ and $(X, Y) = (x, y)$:

$$p(x) = \lim_{\delta x \to 0} \frac{\Pr(x \leq X \leq x + \delta x)}{\delta x}$$
$$p(x, y) = \lim_{\delta x, \delta y \to 0} \frac{\Pr(x \leq X \leq x + \delta x \text{ and } y \leq Y \leq y + \delta y)}{\delta x \delta y}.$$

- Independence if, and only if

$$\Pr(A \cap B) = \Pr(A)\Pr(B).$$

- Total probability theorem for discrete variables, and marginal density of a continuous variable

$$\Pr(X) = \sum_{i=1}^{n} \Pr(X|Y_i)\Pr(Y_i) \qquad p(x) = \int_{-\infty}^{+\infty} p(x|y)p(y)\mathrm{d}y.$$

- Bayes' theorem for transposing conditionals for discrete variables and for continuous variables

$$\Pr(X|Y) = \frac{\Pr(Y|X)\Pr(X)}{\Pr(Y)} \qquad p(x|y) = \frac{p(y|x)p(x)}{p(y)}.$$

- Cumulative distribution function (cdf) for a discrete variable and a continuous a random variable

$$F(x) = \Pr(X \leq x) = \sum_{x_i \leq x} \Pr(x_i) \qquad F(x) = \Pr(X \leq x) = \int_{-\infty}^{x} p(x')\mathrm{d}x'.$$

5

Random variables

> The theory of probabilities is at bottom only common sense reduced to calculus.
>
> Pierre-Simon Laplace,
> *Essai Philosophique sur les Probabilités* (1814)

Random variables are used to model random data and the statistics we calculate from them. In this chapter we shall review some of the properties of random variables, and examine some of the most useful probability distributions.

5.1 Properties of random variables

5.1.1 Expectation

The *expectation* of a discrete random variable X is simply the sum of its possible values $\{x_1, x_2, \ldots, x_n\}$ weighted by their probabilities

$$\mathrm{E}[X] = \sum_{i=1}^{n} x_i p(x_i). \tag{5.1}$$

There is a close connection between this and the sample mean (section 2.3). The corresponding formula for a continuous random variable has the sum replaced by an integral

$$\mathrm{E}[X] = \int_{-\infty}^{+\infty} x p(x) \mathrm{d}x, \tag{5.2}$$

which is the integral of x weighted by the probability density at x. (If the variable has minimum and maximum values x_{min} and x_{max}, these can be used as the limits of the integration.) The expectation value is a measure of the centre of the distribution; we can think of $\mathrm{E}[x]$ as the 'centre of mass' of $p(x)$ imagined as a density spread along the x axis. This is usually called the mean of the distribution (or population mean), and given the special symbol μ.

More generally, if $f(X)$ is some function of the random variable X with density $p(x)$, then the expectation of $f(X)$ is

$$\mathrm{E}[f(X)] = \int_{-\infty}^{+\infty} f(x)p(x)\mathrm{d}x. \tag{5.3}$$

Note that $\mathrm{E}[f(X)]$ is not a function of X since we have integrated over all possible values. The expectation $\mathrm{E}[\cdot]$ is a linear operator. This means that $\mathrm{E}[aX + bY] = a\mathrm{E}[X] + b\mathrm{E}[Y]$, where a and b are constants.

5.1.2 Variance

The *variance* of X is the expectation of the function $(X - \mu)^2$ (where μ is the mean of X), and is denoted $\mathrm{V}[X]$ or given the special symbol σ_x^2 (where the subscript indicates which variable we are referring to). For a discrete random variable, this has the formula

$$\mathrm{V}[X] = \sigma_x^2 = \mathrm{E}[(X - \mu)^2] = \sum_{i=1}^{n}(x_i - \mu)^2 p(x_i). \tag{5.4}$$

As with the mean, there is a close connection between this (sometimes called the population variance) and the sample variance (section 2.4). The corresponding formula for a continuous random variable replaces the sum by an integral,

$$\mathrm{V}[X] = \mathrm{E}[(X - \mu)^2] = \int_{-\infty}^{+\infty} (x - \mu)^2 p(x)\mathrm{d}x. \tag{5.5}$$

Just like the expectation, the variance is not a function of X since we have integrated over all possible values. With a little extra work we can rewrite the variance in a more useful form

$$\begin{aligned}\mathrm{V}[X] &= \mathrm{E}[(X - \mu)^2] = \mathrm{E}[X^2 - 2X\mu + \mu^2]\\ &= \mathrm{E}[X^2] - \mathrm{E}[2X\mu] + \mathrm{E}[\mu^2] = \mathrm{E}[X^2] - 2\mu^2 + \mu^2\\ &= \mathrm{E}[X^2] - \mu^2.\end{aligned} \tag{5.6}$$

This says

variance = expectation of squares − square of expectation.

The variance of a function, $f(X)$, is then

$$\mathrm{V}[f(X)] = \mathrm{E}[f(X)^2] - \mathrm{E}[f(X)]^2. \tag{5.7}$$

The variance gives a measure of the spread of the distribution of X around μ; if the expectation (mean) is the centre of mass of the distribution, then the variance can be thought of as its 'moment of inertia'. Equation 5.6 is then analogous to the parallel axis theorem of mechanics. The variance is often used in theoretical work, and is given the symbol σ^2, but in data analysis it is common to use its positive square root, called the *standard deviation*, given the symbol σ.

The variance is in fact an example of a *moment* of a random variable. The nth moment of the random variable X is defined by $\mu'_n = \mathrm{E}[X^n]$. The nth *central moment* is defined by $\mu_n = \mathrm{E}[(X - \mu)^n]$, in other words the nth moment of $X - \mu$. The mean is the $n = 1$ moment, the variance is the $n = 2$ central moment. Higher-order moments can be important in data analysis, but for most applications, the first two (mean and variance) are the most important.

Box 5.1
Example: rolling dice

Let X be the score from one die roll. Since the six possible outcomes $\{1, 2, 3, 4, 5, 6\}$ are equally probable the expectation is simply

$$\mu = \mathrm{E}[X] = \sum_{i=1}^{6} x_i p(x_i) = \frac{1+2+3+4+5+6}{6} = 3.5.$$

Notice that the expected value is a value than cannot be realised in a single die roll. The variance is

$$\sigma^2 = \sum_{i=1}^{6} (x_i - \mathrm{E}[x])^2 p(x_i) = \frac{1}{6} \sum_{i=1}^{6} (i - 3.5)^2 = \frac{35}{12} = 2.9166\ldots.$$

5.1.3 *Properties of multivariate distributions*

The expectation generalises to cases involving multiple random variables. For example, given the joint density $p(x, y)$ for random variables X and Y, we can define the mean (expectation) of X in the usual way

$$\mu_x = \mathrm{E}[X] = \int_{-\infty}^{+\infty} x p(x) \mathrm{d}x = \int_{-\infty}^{+\infty} x \int_{-\infty}^{+\infty} p(x, y) \mathrm{d}y \mathrm{d}x \qquad (5.8)$$

where we have used the definition of the expectation (equation 5.2) and used equation 4.37 for the marginal distribution $p(x)$. We can find the mean of Y, or the expectation of a function $f(X, Y)$, by the same process. For example, the variance of X

$$\sigma_x^2 = \mathrm{V}[X] = \mathrm{E}[(X - \mu_x)^2] = \int_{-\infty}^{+\infty} (x - \mu_x)^2 \int_{-\infty}^{+\infty} p(x, y) \mathrm{d}y \mathrm{d}x \qquad (5.9)$$

and the expectation of $X + Y$ is

$$\mathrm{E}[aX + bY] = a\mathrm{E}[X] + b\mathrm{E}[Y]. \tag{5.10}$$

Box 5.2

Proof of $\mathrm{E}[aX + bY] = a\mathrm{E}[X] + b\mathrm{E}[Y]$

For continuous variables X and Y, and constants a and b, we have

$$\begin{aligned}
\mathrm{E}[aX + bY] &= \iint (ax + by)p(x, y)\mathrm{d}x\mathrm{d}y \\
&= a\int x \int p(x, y)\mathrm{d}y\mathrm{d}x + b\int y \int p(x, y)\mathrm{d}x\mathrm{d}y \\
&= a\int xp(x)\mathrm{d}x + b\int yp(y)\mathrm{d}y \\
&= a\mathrm{E}[X] + b\mathrm{E}[Y]
\end{aligned}$$

where we made use of equation 4.37 in the third line, and equation 5.8 in the fourth line. This result holds whether or not the variables are independent.

Similarly, for two functions $f(X)$ and $g(Y)$ we have that

$$\mathrm{E}[af(X) + bg(Y) + c] = a\mathrm{E}[f(X)] + b\mathrm{E}[g(Y)] + C. \tag{5.11}$$

Given the joint density for two random variables we can also define their *covariance*, given the special symbol σ_{xy}:

$$\begin{aligned}
\mathrm{Cov}(X, Y) = \sigma_{xy} &= \mathrm{E}[(X - \mu_x)(Y - \mu_y)] \\
&= \mathrm{E}[XY - X\mu_y - Y\mu_x + \mu_x\mu_y] \\
&= \mathrm{E}[XY] - \mathrm{E}[X\mu_y] - \mathrm{E}[Y\mu_x] + \mathrm{E}[\mu_x\mu_y] \\
&= \mathrm{E}[XY] - \mu_x\mu_y
\end{aligned} \tag{5.12}$$

(compare with equation 5.6). Note that the covariance of X and X is the variance of X: $\sigma_{xx} = \sigma_x^2$. Another important relation between covariance and variance is

$$|\sigma_{xy}|^2 \le \sigma_x^2\sigma_y^2 \quad \text{or} \quad |\mathrm{Cov}(X, Y)|^2 \le \mathrm{V}[X]\mathrm{V}[Y] \tag{5.13}$$

known as the Cauchy–Schwarz inequality. The *correlation coefficient* is the covariance of two variables, scaled by their standard deviations

$$\rho(X, Y) = \frac{\sigma_{xy}}{\sigma_x\sigma_y}. \tag{5.14}$$

These can be compared to the sample covariance and correlation coefficient (section 2.8). The correlation coefficient ranges from between -1 and $+1$ (see equation 5.13).

The expectation of the product of two independent variables x and y is the product of their individual expectations

$$\begin{aligned}
\mathrm{E}[XY] &= \int_{-\infty}^{+\infty} xyp(x, y)\mathrm{d}x\mathrm{d}y \\
&= \left[\int_{-\infty}^{+\infty} xp(x)\mathrm{d}x\right]\left[\int_{-\infty}^{+\infty} yp(y)\mathrm{d}y\right] = \mu_x\mu_y. \qquad (5.15)
\end{aligned}$$

(This is not generally true when the variables are not independent, i.e. when equation 4.40 does not hold.) Comparing this result with equations 5.12 we can see that the covariance (and correlation coefficient) will vanish for mutually independent variables.

These ideas can be extended to cases with several random variables. When there are more than two variables, e.g. $\{X_1, X_2, \ldots, X_n\}$, we can define the covariances between each pair of variables, σ_{ij}^2, and assemble these in a square $(n \times n)$ matrix called the *covariance matrix*. In the three-dimensional case, for variables X, Y and Z, the covariance matrix will look like

$$\Sigma = \begin{pmatrix} \sigma_{xx} & \sigma_{xy} & \sigma_{xz} \\ \sigma_{yx} & \sigma_{yy} & \sigma_{yz} \\ \sigma_{zx} & \sigma_{zy} & \sigma_{zz} \end{pmatrix}. \qquad (5.16)$$

The elements along the leading diagonal (top left–bottom right) are the variances σ_x^2, σ_y^2 and σ_z^2. Notice also that this matrix is symmetric about the leading diagonal, since $\sigma_{xy} = \sigma_{yx}$.

5.1.4 Linear functions of random variables

The simplest combination of several variables is a linear combination. The mean of a set of random variables is an example of a linear combination, so now we shall turn to the properties of linear combinations.

We write a linear function of several random variables $\{X_1, X_2, \ldots, X_n\}$, using coefficients a_i, as

$$Y = a_1X_1 + a_2X_2 + \cdots a_nX_n = \sum_{i=1}^{n} a_iX_i. \qquad (5.17)$$

The expectation of this linear function is then

$$\mathrm{E}[Y] = E\left[\sum_{i=1}^{n} a_i X_i\right] = \sum_{i=1}^{n} a_i \mathrm{E}[X_i] = \sum_{i=1}^{n} a_i \mu_i \tag{5.18}$$

which is an extension of equation 5.10 (and has a similar proof). In words this says

expectation of a weighted sum = weighted sum of expectations.

The variance of the linear combination Y is

$$\mathrm{V}[Y] = \sum_{i=1}^{n} a_i^2 \sigma_i^2 + \sum_{i=1}^{n} \sum_{j \neq i} a_i a_j \sigma_{ij}^2. \tag{5.19}$$

(To get to this expression one has to expand the expression for the variance, equation 5.5, and follow some long-winded algebra. See e.g. section 4.7 of Miller and Miller 2003.) If the X_i are mutually uncorrelated, then $\sigma_{ij}^2 = 0$ for all $i \neq j$, the second term in the above equation vanishes, and we are left with

$$\mathrm{V}[Y] = \sum_{i=1}^{n} a_i^2 \sigma_i^2. \tag{5.20}$$

Therefore, the variance of a linear function (of independent variables) is a quadratic function in the expansion coefficients a_i.

We can use the above relations to compute the expectation and variance of a sample mean. We start with n random variables $\{X_1, X_2, \ldots, X_n\}$ all with the same mean and variance (μ and σ^2), and form a new random variable from the mean of these n variables

$$\bar{X} = \frac{1}{n} \sum_{i=1}^{n} X_i. \tag{5.21}$$

We could, for example, be taking the mean height of groups of $n = 20$ people from a population. The variables X_i are the heights of individual people, and $\bar{X}$ is another random variable formed from averages of 20 individual heights. In this case μ and σ^2 are the population mean and variance. The sample mean (above) is just a linear combination (equation 5.17) with coefficients $a_i = 1/n$. We can therefore write down its expectation

$$\mathrm{E}[\bar{X}] = \frac{1}{n} \sum_{i=1}^{n} \mathrm{E}[X_i] = \frac{1}{n} \sum_{i=1}^{n} \mu = \frac{1}{n}(n\mu) = \mu, \tag{5.22}$$

which says that the expectation of the sample mean is the population mean (as we should expect). If the variables X_i are mutually independent, the variance of the sample mean is given by 5.20:

$$\mathrm{V}[\bar{X}] = \frac{1}{n}\sigma^2 \quad \Rightarrow \quad \sigma_{\bar{x}} = \sqrt{\mathrm{V}[\bar{X}]} = \frac{\sigma}{\sqrt{n}} \tag{5.23}$$

which says that the standard deviation of the mean scales as $n^{-1/2}$ (under the stated assumptions). This is the argument that underlies the standard error of equation 2.4.

5.2 Discrete random variables

So far we have dealt with many rules for combining and manipulating probabilities and probability densities. These are like the grammar of the language of probability. But we have not yet covered the functions that describe these probabilities and densities, which are like the basic vocabulary. These are used to build probability models for random process, data and statistics. In the following few sections, we briefly examine a few of the most important distribution functions.

Our starting point will be a simple example: taking sweets from a bag containing a mixture of red and green sweets. First, imagine we have just one sweet of each colour in the bag, and pull a sweet out of the bag 'at random', look at its colour and then return it to the bag. This is an example of *sampling with replacement*. We can form a probability model of this situation by assigning probabilities Pr(*red*) = 1 − Pr(*green*) = 0.5. But we could try a similar experiment with three red sweets and two green sweets, and model this as Pr(*red*) = 1 − Pr(*green*) = 0.6. We do this by assigning equal probabilities to each sweet. But we are only interested in the colour of the sweet, not its individual identity (we could have given a unique number to each sweet), and so any of the three red sweets counts as *red*. This is an example of a *Bernoulli trial*.[1]

We can write down the probability function for the Bernoulli trial as

$$p(x) = \begin{cases} \theta & \text{for } x = 1 \\ 1 - \theta & \text{for } x = 0. \end{cases} \tag{5.24}$$

Here X is a random variable with two possible values $x \in \{0, 1\}$, and θ is the probability at $x = 1$. In our sweets example we could have X the random variable that takes value 1 when a red sweet is drawn, and 0 when a green sweet is drawn, and θ is the probability of drawing a red sweet (given by the fraction of red sweets in the bag). This can be written more compactly as the *Bernoulli distribution* function

$$p(x|\theta) = \theta^x (1 - \theta)^{1-x} \quad \text{for } x = 0, 1. \tag{5.25}$$

[1] Named after the Swiss mathematician Jacob Bernoulli (1654–1705). He wrote *Ars Conjectandi* (published posthumously), often consider the first serious mathematical treatment of probability theory. He also discovered the mathematical constant e.

Notice the way the notation runs: $p(x|\theta)$ means the probability (mass or density) as a function of x (the different values that random variable may take), given a specific value of the parameter θ. The function defines a whole family of distributions, one for each value of θ. Once θ is known we know which member of the family to use. It is traditional to name the possible outcomes of a Bernoulli trial *success* and *failure*, but the names do not matter, they could be *red* and *green*, or 1 and 0.

If a random variable X follows the Bernoulli distribution, we write

$$X \sim \text{Bern}(\theta)$$

where the tilde '$\sim$' means 'is distributed as' and we use here Bern(θ) to indicate the Bernoulli distribution with parameter θ. This is standard shorthand in mathematical statistics.

We can use equations 5.1 and 5.4 to compute the expectation and variance of a Bernoulli distributed variable. Substitution of the probabilities gives $\text{E}[X] = \theta$ and $\text{V}[X] = \theta(1 - \theta)$.

5.2.1 Binomial distribution

The Bernoulli model is useful for describing simple situations, but more often we will be faced with something more like repeated Bernoulli trials. That is, we repeat an experiment, which has two possible outcomes, n times, and each time the conditions are essentially the same. If the probability of each outcome does not change from one trial to the next, then each is an independent Bernoulli trial. In this case we need the probability function that describes the probability of x successes from n trials. If the probability of drawing a red sweet from the bag is θ, what is the probability that we draw x red sweets if we sample (with replacement) from the bag n times?

Box 5.3
Binomial coefficient

The number of ways of arranging x successes among n observations is given by the *binomial coefficient*

$$\binom{n}{x} = \frac{n!}{(n-x)!x!} \tag{5.26}$$

for $x = 0, 1, 2, \ldots, n$. This is often pronounced 'n choose x'.

For example, with $x = 2$ red sweets and 1 green sweet, making a total of $n = 3$ sweets, we have

$$\binom{3}{2} = \frac{3!}{(3-2)!2!} = \frac{6}{2} = 3 \tag{5.27}$$

arrangements. These are $\{r, r, g\}$, $\{r, g, r\}$, $\{g, r, r\}$. See Appendix C for a brief reminder of combinations and permutations.

Let's start simple. We make $n = 3$ draws and find $x = 2$ reds (and therefore $n - x = 1$ green). What is the probability for this? The probability of drawing two reds and one green, in that order, is

$$\Pr(red)\Pr(red)\Pr(green) = \theta\theta(1-\theta) = \theta^2(1-\theta)$$

using the multiplication rule for independent events (equation 4.11). But we are not interested in the order of the reds, only their number. And there are three ways to arrange two reds and a green ($\{r, r, g\}$, $\{r, g, r\}$, $\{g, r, r\}$), each of which is equally probable. (See box 5.3 for the number of ways to arrange events.) We therefore sum the probabilities for each arrangement (using the addition rule for independent events, equation 4.9) to get the probability of obtaining exactly two reds ($x = 2$) from three draws ($n = 3$) when the order does not matter:

$$\Pr(2\ red \text{ and } 1\ green) = 3\theta^2(1-\theta)^1.$$

R.Box 5.1

Simulating repeated Bernoulli trials

If we have three red and two green sweets in a bag and draw (with replacement) five times, what is the probability of getting zero reds? Or one red only? We can simulate this process (see also R.box 4.1) using the following code fragment.

```
Omega <- c("red", "green")
n <- 5
N.sims <- 1000
X.sim <- array(0, dim=N.sims)
for (i in 1:N.sims) {
  samp <- sample(Omega, prob=c(0.6, 0.4), size=n,
                 replace=TRUE)
  X.sim[i] <- sum(samp == "red")
}
```

The first line defines the sample space (`Omega`) in which we put the sweets. We then use `n` to define how large each sample is (e.g. five draws from the bag), and `N.sims` to define how many times to repeat the simulation to build up a distribution. The array `X.sim` is made ready to store the results of each simulation. The next few lines define a loop that repeats the random sampling (the lines between curly brackets) once for each `i=1,2,...,N.sims`, and stores the number of reds in `X.sim[i]`. Once this has finished we plot a histogram, and store the histogram data in a new object, `h`, for later use.

```
breaks <- (-1:n) + 0.5
h <- hist(X.sim, breaks=breaks)
```

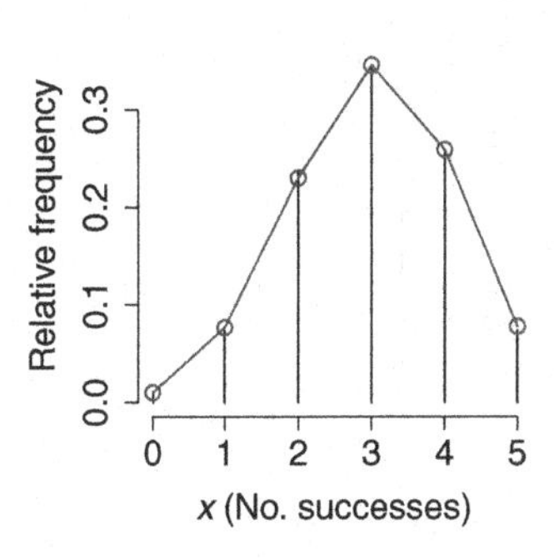

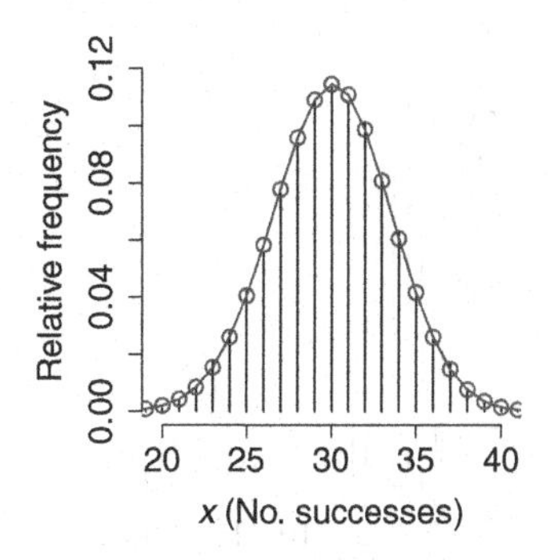

Figure 5.1 The bar chart data show the results of 10 000 simulations of random sampling. Each simulation gives a random number of 'successes' in n trials (where $n = 5$ on the left, and $n = 50$ on the right), where the probability of 'success' (per trial) is $\theta = 0.6$. The hollow circles (joined by lines) show the corresponding exact binomial distribution.

R.Box 5.2
The binomial distribution emerges

We can compare the simulation results from R.box 5.1 with the binomial distribution. We first make a neater bar chart showing the relative frequencies for drawing $0, 1, \ldots, n$ sweets from our simulations. The histogram object `h` contains arrays `mids` and `density` that list the mid-points of the histogram bins, and the density (i.e. the frequency relative to the total number of simulations).

```
plot(h$mids, h$density, type="h", lwd=2,
     xlab="x (No. reds)", ylab="Relative frequency")
```

We can compare this with the binomial probabilities:

```
p.binom <- dbinom(h$mids, size=n, prob=0.6)
lines(h$mids, p.binom, type="o", col="red")
```

Now try repeating the simulation process using a sample size $n = 50$ and compare this with the appropriate binomial distribution. Figure 5.1 shows some example results.

R.Box 5.3
Probability distributions

R has a wide range of probability distributions included, and is capable of calculating their density (or mass) function, cumulative probability function and quantiles, and also generating random numbers. The commands for each type of distribution have a

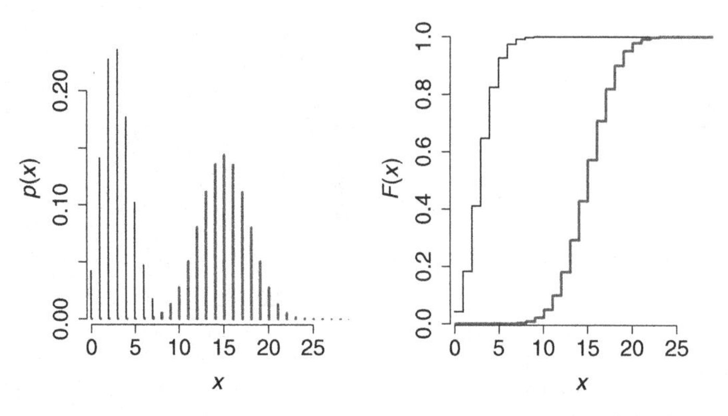

Figure 5.2 Probability mass function (pmf; left) and cumulative mass function (cmf; right) for the binomial distribution, with $n = 30$ and $\theta = 0.1$ (black) and $\theta = 0.5$ (thick grey).

similar form. For a probability (density) the function begins with `d` and for random numbers it begins with `r`. For example, the normal distribution can be used as follows:

```
dnorm(0.2)          # density p(x) at x=0.2
pnorm(0.2)          # cumulative Pr(<x)
qnorm(0.5)          # find 0.5 quantile (median)
rnorm(20)           # 20 random numbers
```

In the general case, the probability of drawing x red sweets if we make n repeated draws, each with probability θ of drawing a red, is given by the *binomial distribution*,[2] one of the most important distributions in statistics:

$$\begin{aligned} p(x|n,\theta) &= \begin{pmatrix} \text{number of} \\ \text{matching sequences} \end{pmatrix} \begin{pmatrix} \text{probability of any} \\ \text{particular matching sequence} \end{pmatrix} \\ &= \binom{n}{x} \theta^x (1-\theta)^{n-x} = \frac{n!}{(n-x)!x!} \theta^x (1-\theta)^{n-x} \end{aligned} \tag{5.28}$$

where $x = 0, 1, 2, \ldots, n$. The second part of the formula, giving the probability of each combination, is a generalisation of the Bernoulli distribution (equation 5.25). The Bernoulli distribution is a special case of the binomial with $n = 1$. Figure 5.2

[2] The binomial distribution gets its name from its similarity to the binomial expansion. The terms $p(x|n, \theta)$ for $x = 0, 1, 2, \ldots, n$ are the same as the successive terms in the binomial expansion of $[\theta + (1-\theta)]^n$. This fact is enough to show the probability summed over all x values, $\sum_{x=0}^{n} p(x|n, \theta) = 1$.

shows some examples of binomial probabilities. If a random variable X follows this distribution we usually write

$$X \sim \text{Binom}(n, \theta).$$

We can find the expectation and variance of the binomial distribution by comparing it the Bernoulli distribution.[3] A binomial distribution represents repeated Bernoulli trials, so we can form a binomial variable from the sum of n Bernoulli variables, $X = X_1 + X_2 + \cdots X_n$, where the $X_i \sim \text{Bern}(\theta)$ and so have $\text{E}[X_i] = \theta$ and $\text{V}[X_i] = \theta(1 - \theta)$. Then applying what we know about linear combinations of random variables (equations 5.18 and 5.20) we find $E[X] = n\theta$ and $\text{V}[X] = n\theta(1 - \theta)$. It is also straightforward to show that the sum of two binomial variables follows a binomial distribution, so if $X \sim \text{Binom}(m, \theta)$ and $Y \sim \text{Binom}(n, \theta)$ then $Z = X + Y \sim \text{Binom}(m + n, \theta)$.

Many other distribution functions are related to the binomial. The *multinomial* is an extension to allow for more than two outcomes per trial. The *negative binomial* gives the probability for the kth success in a series of Bernoulli trials. The *hypergeometric distribution* applies to sampling without replacement (recall that the binomial distribution is a model for sampling with replacement). See e.g. Miller and Miller (2003) or Casella and Berger (2001) for more details of these.

Box 5.4

Example: Throwing many dice

What is the probability of rolling no sixes from six rolls of a die? The probability of a six from one trial is $1/6$ so

$$p(x = 0|n = 6, \theta = 1/6) = \frac{6!}{0!6!}(1/6)^0(5/6)^6 = \left(\frac{5}{6}\right)^6 = 0.335.$$

What is the probability of rolling at least one six from six rolls? This is simply $1 - p(0) = 0.665$ Finally, what is the probability of rolling exactly two sixes from six rolls?

$$p(x = 2|n = 6, \theta = 1/6) = \frac{6!}{2!4!}(1/6)^2(5/6)^4 = 0.201.$$

[3] It is also possible to derive this result by analysis of the probability function, see e.g. Miller and Miller, 2003, section 5.4, or Casella and Berger, 2001, section 2.3.

R.Box 5.4

The binomial distribution in R

The binomial distribution $p(x|n, \theta)$ may be visualised for any particular n and θ by using the `dbinom()` function to calculate the probability function for the binomial distribution, and then plotting a bar chart, as follows:

```
x <- 0:20
p <- dbinom(x, size=30, prob=0.1)
plot(x, p, xlab="x", ylab="p(x)", type="h")
```

The first line generates a vector of x values $\mathbf{x} = \{0, 1, 2, \ldots, 20\}$. The second line generates a vector containing the values of the binomial probability function at each of the corresponding x values. In this case we choose $n = 30$ and $\theta = 0.1$. The final line plots $p(x)$ at each x as a bar chart (specified with the `type="h"` argument).

R.Box 5.5

Random numbers from the binomial distribution

Random numbers following a binomial distribution can be generated using the `rbinom()` function as follows:

```
y <- rbinom(1000, size=30, prob=0.1)
plot(y)
hist(y, breaks=x-0.5, prob=TRUE, col="blue")
lines(x, p, lwd=4)
```

The first line generates 1000 random numbers from the binomial distribution with $n = 30$ and $\theta = 0.1$. The second line plots the resulting sequence. The third line plots a histogram of the 1000 random values. The `breaks=x-0.5` positions the breaks between histogram bins, and the `prob=TRUE` argument plots the histogram as a probability density rather than total absolute frequency. The last line compares the histogram of 1000 random numbers with the exact binomial distribution calculated previously.

Box 5.5

Example: How many trials to succeed?

Suppose some desired event has a probability p of occurring in each trial. How many trials must be performed in order for the probability of at least one event happening to reach α? If x is the number of events in n trials, we wish to find x such that

$$p(x \geq 1) \geq \alpha \quad \text{or} \quad 1 - p(x = 0) \geq \alpha \quad \text{or} \quad p(x = 0) \leq 1 - \alpha.$$

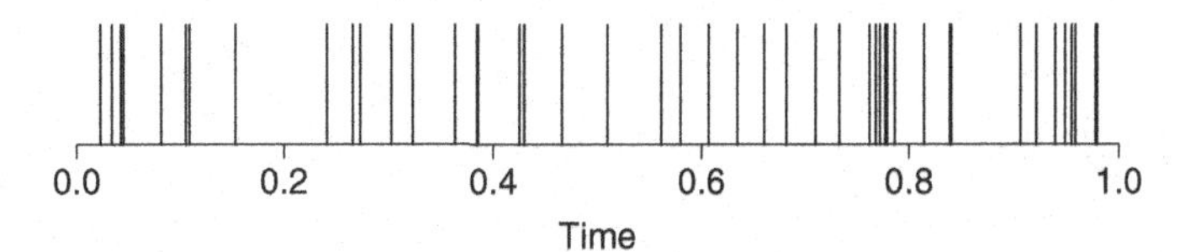

Figure 5.3 A sequence of randomly timed pulses.

Using the expression for the binomial distribution with $x = 0$ in the rightmost inequality we get

$$(1-\theta)^n \leq 1-\alpha.$$

We can take the log of both sides and rearrange (noting that $1-\theta < 1$, so its logarithm is negative) to find

$$n \geq \frac{\log(1-\alpha)}{\log(1-\theta)}.$$

For example, imagine that the probability of passing your driving test is $\theta = 0.5$. How many tests should you be prepared to take for the probability of passing one of them to exceed 0.99? Following the above analysis we have $n \geq \log(1-0.99)/\log(1-0.5) \approx 6.6$.

5.2.2 Poisson distribution

The *Poisson distribution* is related to the binomial distribution (see box 5.6), and is very important in several areas of physics, notably particle physics. Particle detection is similar to a Bernoulli trial: either the particle registers in the detector or it does not. But instead of there being a fixed number of trials, the detections may occur at any point within a continuous time interval. Figure 5.3 shows a sequence of pulses that occur randomly in time – if these occur at a fixed average rate (in the 'long run'), but the actual timings of each event are random and independent of each other, then the number of events over a finite interval will follow a Poisson distribution. This is written

$$p(x|\lambda) = \frac{\lambda^x e^{-\lambda}}{x!} \tag{5.29}$$

where $x = 0, 1, 2, \ldots$ (this is the *support* of the distribution) and λ is the parameter that sets the expected rate, and can be any positive real ($\lambda > 0$). If the random variable X follows a Poisson distribution, we say

$$X \sim \text{Pois}(\lambda).$$

Figure 5.4 shows some example Poisson distributions. The expectation and variance of the Poisson distribution are $E[X] = \lambda$ and $V[X] = \lambda$, which can be seen by

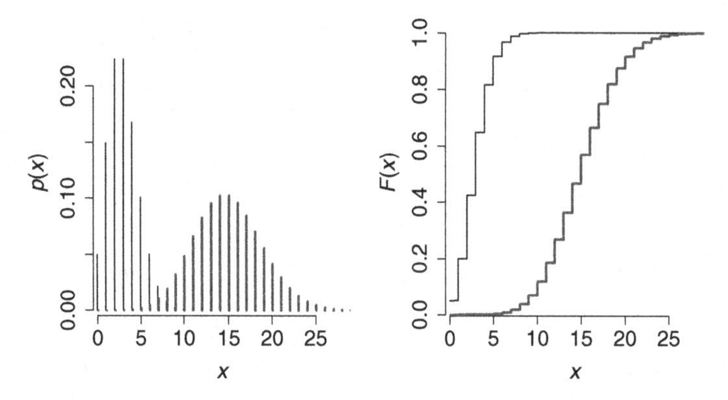

Figure 5.4 Probability mass function (pmf; left) and cumulative mass function (cmf; right) for the Poisson distribution with $\lambda = 3$ (black) and $\lambda = 15$ (thick grey).

applying the definitions of equations 5.1 and 5.4. For example,

$$\begin{aligned} \mathrm{E}[X] &= \sum_{x=0}^{\infty} x p(x) = \sum_{x=0}^{\infty} x \frac{\lambda^x \mathrm{e}^{-\lambda}}{x!} = \lambda \mathrm{e}^{-\lambda} \sum_{x=1}^{\infty} \frac{\lambda^{x-1}}{(x-1)!} \\ &= \lambda \mathrm{e}^{-\lambda} \left[\sum_{y=0}^{\infty} \frac{\lambda^y}{(y)!} \right] = \lambda \mathrm{e}^{-\lambda} [\mathrm{e}^{\lambda}] = \lambda. \end{aligned}$$

In the first line we have used $x/x! = 1/(x-1)!$ and dropped the $x = 0$ term from the summation (which adds zero). In the second line we made the substitution $y = x - 1$, then used the Taylor series expansion $\mathrm{e}^{\lambda} = \sum_{y=0}^{\infty} \lambda^y / y!$. A similar analysis gives the variance $\mathrm{V}[X] = \lambda$.

R.Box 5.6

The Poisson distribution in R

The functions for using the Poisson distribution are similar to those for the binomial distribution. For example, to calculate the probability of three events in a day, from an average rate of $\lambda = 1.5$ per day, use

```
dpois(3, lambda=1.5)
```

and to find the probability of ≤ 3 per day use

```
ppois(3, lambda=1.5)
```

which calculates the cumulative probability up to (and including) 3.

Box 5.6
Relationship between Poisson and binomial distributions

One way to obtain the Poisson distribution is as a limiting case of the binomial distribution for $n \to \infty$ and $\theta \to 0$ such that their product $n\theta = \lambda$ is finite (the expectation for a binomial distribution). We can write the binomial probability mass function using $\theta = \lambda/n$:

$$p(x|n, \lambda/n) = \frac{1}{x!} \frac{n!}{(n-x)!} \frac{\lambda^x}{n^x} \left(1 - \frac{\lambda}{n}\right)^{n-x},$$

and consider the limit as $n \to \infty$. The first term stays the same in the limit. In the limit we have $x \ll n$ and so the second term, $(n)(n-1)\cdots(n-x+1) \to n^x$. The product of this with the third term gives λ^x. The fourth term can be approximated

$$\lim_{n\to\infty} (1 - \lambda/n)^{n-x} = \lim_{n\to\infty} (1 - \lambda/n)^n = (e^{-1})^\lambda = e^{-\lambda}$$

(using the definition of $e^x = \lim_{n\to\infty}(1 + x/n)^n$). Combining these terms we get the Poisson probability mass function (equation 5.29).

R.Box 5.7
Random numbers from the Poisson distribution

To generate random numbers from the Poisson distribution, use the `rpois()` command. For example, to generate 100 numbers from a distribution with a mean of $\lambda = 3.5$, plot their histogram and overlay the Poisson curve, use

```
x <- rpois(100, lambda=3.5)
plot(x)
breaks <- -1:12 + 0.5
h <- hist(x, breaks=breaks, prob=TRUE, col="grey")
pmf.pois <- dpois(h$mids, lambda=3.5)
lines(h$mids, pmf.pois, type="o")
```

How does the plot change with 5000 random numbers?

Box 5.7
Example: How many gamma-ray bursts?

Suppose the mean rate of cosmic gamma-ray bursts (GRBs) is 1.5 per day. What is the probability for getting more than two bursts in any given day? The Poisson probabilities for $x = 0, 1, 2$ assuming an expected rate of $\lambda = 1.5$ are

$$p(0|\lambda = 1.5) = 1.5^0 e^{-1.5}/0! = 0.223$$

$$p(1|\lambda = 1.5) = 1.5^1 e^{-1.5}/1! = 0.335$$

$$p(2|\lambda = 1.5) = 1.5^2 e^{-1.5}/2! = 0.251.$$

The probability of a day with ≥ 3 GRBs is $p(x \geq 3|\lambda = 1.5) = 1 - [p(0) + p(1) + p(2)] = 0.191$.

The Poisson distribution recurs throughout science as a useful description of discrete phenomena occurring randomly and independently over an interval (at a fixed average rate). It describes a *counting* process. Here are some examples of processes we can model using the Poisson function.

- Number of photons detected per pixel while taking an image.
- The number of stars per volume of space.
- Number of mutations on certain length of DNA upon exposure to a fixed dose of radiation.
- Number of emergency calls at a given hour of the day.

5.3 Continuous random variables

5.3.1 Normal distribution

The *normal*, or *Gaussian*, distribution is (probably) the most important distribution in statistics. This is the distribution with the famous 'bell-shaped curve'. A continuous random variable X has a normal distribution if its density is of the form

$$p(x|\mu, \sigma^2) = \frac{1}{\sigma\sqrt{2\pi}} e^{-(x-\mu)^2/(2\sigma^2)}. \tag{5.30}$$

The normal distribution is specified by two parameters, the mean μ and variance σ^2. These define the centre (location) and width (scale) of the distribution. Figure 5.5 shows a normal distribution and the probability mass enclosed in various ranges about the mean. Figure 4.5 illustrates the cumulative and density functions for the normal distribution. If a random variable X has a normal distribution, we write

$$X \sim N(\mu, \sigma^2),$$

and the variable $Z = (X - \mu)/\sigma$ has an $N(0, 1)$ distribution, known as the *standard normal*, which has density

$$p(z|0, 1) = \frac{1}{\sqrt{2\pi}} e^{-z^2/2}. \tag{5.31}$$

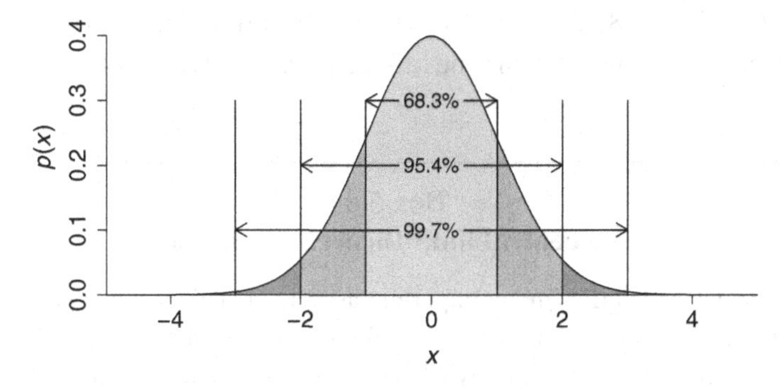

Figure 5.5 Standard normal pdf with the 1σ, 2σ and 3σ ranges (and enclosed probabilities) indicated.

This formula does not have an anti-derivative that can be written in terms of elementary functions, so integrating the normal density function requires special treatment (in general we use numerical methods to evaluate the definite integral). It turns out that the expectation and variance for the standard normal are $\mathrm{E}[Z] = 0$ and $\mathrm{V}[Z] = 1$. That the expectation is zero should be obvious once we notice that the density is symmetric about zero (the half-integrals either side of zero cancel exactly). (See e.g. section 6.5 of Miller and Miller (2003) or 3.3 of Casella and Berger (2001) for details.) The expectation and variance of a (non-standard) normal are then $\mathrm{E}[X] = \mu$ and $\mathrm{V}[X] = \sigma^2$, by application of equations 5.11 and 5.20 to $X = Z\sigma + \mu$.

The *central limit theorem* (CLT) is a powerful theorem of central importance in mathematical statistics and practical data analysis. If we have a sequence of independent random variables X_i, each from a distribution with mean μ_i and variance σ_i^2, then the distribution of the sum $Y = \sum X_i$ will have a mean $\sum \mu_i$ and a variance $\sum \sigma_i^2$ (section 5.1.4). The CLT tells us about the distribution of Y: the distribution of Y becomes more like a normal as n increases (under quite general conditions, such as that the means μ_i and variances σ_i^2 are finite). This is a very powerful theorem; it means that the sum or average of a sufficient number of independent random data will follow a normal distribution, even if the distribution of the original data is not normal or even unknown. (For mathematical details, see e.g. section 8.2 of Miller and Miller (2003) or section 5.5 of Casella and Berger (2001).)

It is also true – for similar reasons – that other distributions converge on the normal in their limits. The binomial converges on a normal distribution as $n \to \infty$ (see Figure 5.2), as does the Poisson distribution as $\lambda \to \infty$ (see Figure 5.4).

This explains why the normal distribution is so useful and important: the cumulative effect of a large number of independent events tends to be normally distributed (see box 5.8). Like many other distributions, the normal distribution may be extended to more than one dimension, but we shall not discuss this here.

Box 5.8

The central limit theorem in physics

In physics the CLT underlies many important aspects of statistical mechanics. For example, gas molecules randomly colliding (elastically) with each other transfer momentum: each collision effectively adds or subtracts a random amount of momentum. The momentum along the x-axis is the sum of all the x-direction momentum transfers from all the previous collisions. The CLT says that after a large number of collisions the distribution of the x-component of momentum (and hence the x-component of velocity) will tend to normal. The same is true in the other directions.

R.Box 5.8

The normal distribution in R

Let's generate 500 random numbers using a normal distribution with mean $\mu = 1$ and variance $\sigma^2 = 4$ ($\sigma = 2$) and compare these to the normal density:

```
x <- rnorm(5000, mean=1, sd=2)
plot(x)
hist(x, prob=TRUE, col="grey")
xx <- seq(-10, 10, by=0.1)
pdf.norm <- dnorm(xx, mean=1, sd=2)
lines(xx, pdf.norm)
```

5.3.2 Chi-square distribution

Suppose that $X_1, X_2, \ldots, X_n$ are independent, standard normal variables, with distribution $N(0, 1)$. The sum of their squares, which is itself a random variable, $X = \sum_{i=1}^{n} X_i^2$, is said to have a χ^2 (*chi-square*) distribution. The χ^2 distribution is continuous in its variable, which is always positive (since the X_i^2 are positive). It is commonly encountered when one is dealing with the sum of squares of normally distributed data. Recall that the sample variance is the sum of squared deviations (equation 2.3), so this often has a χ^2 distribution. (In fact $(n-1)s^2/\sigma^2$ has a χ^2 distribution with $\nu = n - 1$ degrees of freedom.) This fact will be useful when we discuss goodness-of-fit tests later on.

The continuous random variable X has a χ^2 distribution if its density is of the form[4]

$$p(x|\nu) = \frac{(1/2)^{\nu/2}}{\Gamma(\nu/2)} x^{\nu/2-1} e^{-x/2} \tag{5.32}$$

for $x \geq 0$. The parameter ν (=1, 2, ...) is called the 'degrees of freedom'. If X is a sum of squares of n normal variables, it follows a χ^2 distribution with $\nu = n$ degrees of freedom, which we write as

$$X \sim \chi^2_\nu.$$

Notice that in this case the parameter (ν) is (by tradition) written as a subscript. The expectation and variance of the χ^2 distribution are $\mathrm{E}[X] = \nu$ and $\mathrm{V}[X] = 2\nu$ (which we state without proof).

Box 5.9
The χ^2 distribution in physics

Recall that the distributions of the x, y and z components of the momentum for gas molecules (in thermal equilibrium) each follow a normal distribution (and are independent, since the axes are orthogonal) by the central limit theorem, as a result of the very large number of collisions with other molecules. The kinetic energy of a molecule is proportional to the square of the total velocity ($E = \frac{1}{2}mv^2$) and by Pythagoras's theorem $v^2 = v_x^2 + v_y^2 + v_z^2$. So the kinetic energy of a molecule is proportional to the sum of the squares of the velocities in the x, y and z directions.

As each velocity component is normally distributed, the sum of their squares is χ^2 distributed, with $\nu = 3$ degrees of freedom. Hence the kinetic energy for molecules of a gas in thermal equilibrium has a χ^2_3 distribution. In fact $X = 2E/kT \sim \chi^2_3$ (where E, k and T have their usual thermodynamics meanings). This is connected to the Maxwell–Boltzmann distribution for the distribution of molecular speeds (which is a 'non-central chi distribution' with three degrees of freedom).

If X_n^2 and X_m^2 have independent χ^2 distributions with n and m degrees for freedom, respectively, the sum $X_k^2 = X_m^2 + X_n^2$ has a χ^2 distribution with $k = n + m$ degrees for freedom. This should be obvious from the definition above. Asymptotically, the χ^2_ν distribution becomes normal as $\nu \to \infty$. At small ν the distribution is asymmetric (see Figure 5.6).

[4] The Gamma function, $\Gamma(\cdot)$ is an extension of the factorial function to continuous variables, and is defined by the integral

$$\Gamma(z) = \int_0^\infty t^{z-1} e^{-t} dt.$$

When the argument is a positive integer, the Gamma function is the familiar factorial function but offset by one: $\Gamma(n) = (n-1)!$. The first few useful values are $\Gamma(1/2) = \sqrt{\pi}$, $\Gamma(1) = 1$, $\Gamma(3/2) = \sqrt{\pi}/2$, $\Gamma(2) = 1$.

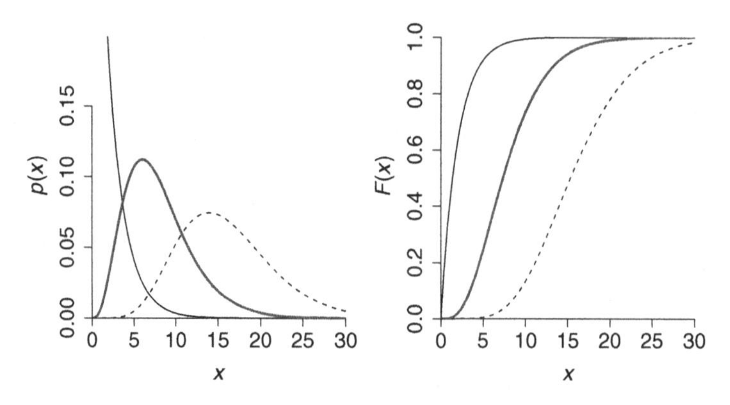

Figure 5.6 Probability density function (pdf; left) and cumulative density functions (cmf; right) for members of the family of chi-square pdfs: $\nu = 2$, $\nu = 8$ and $\nu = 16$ (black, thick grey and dashed, respectively). The distributions are more symmetric (more like a normal curve) for higher ν.

R.Box 5.9

The χ^2 distribution in R

We can generate random numbers with a χ^2 distribution as follows:

```
x <- rchisq(500, df=5)
plot(x)
h <- hist(x, prob=TRUE, col="grey")
xx <- seq(0, 100, by=0.1)
pdf.chisq <- dchisq(xx, df=5)
lines(xx, pdf.chisq)
```

Now, repeat this using $\nu = 100$. The distribution should look more like a normal one.

R.Box 5.10

Adding χ^2 distributed data

We can generate two sets of random numbers, x and y, each set from a different χ^2 distribution, and examine the distribution of the sum $z = x + y$, which should itself follow a χ^2 distribution

```
x <- rchisq(500, df=5)
y <- rchisq(500, df=4)
z <- x + y
h <- hist(z, prob=TRUE, col="grey")
xx <- seq(0, 100, by=0.1)
pdf.chisq <- dchisq(xx, df=9)
lines(xx, pdf.chisq)
```

The density function for the χ^2 distribution with $\nu = 1$ is

$$p(x) = \frac{1}{\sqrt{2\pi}} x^{-1/2} \mathrm{e}^{-x/2}, \tag{5.33}$$

and with $\nu = 2$ we have an exponential distribution

$$p(x) = \frac{1}{2} \mathrm{e}^{-x/2}. \tag{5.34}$$

The exponential distribution is familiar to physicists as the distribution of decay times for radioactive particles. As $\nu \to \infty$ the χ^2_ν distribution approaches normal with mean ν and variance 2ν: $N(\nu, 2\nu)$.

5.3.3 Student's t-distribution

If $X_1, X_2, \ldots, X_n$ are normally distributed with mean μ and variance σ^2, then their mean $\bar{X}$ is also normally distributed: $N(\mu, \sigma^2/n)$ (see section 5.1.4). We can *estimate* the variance of the n variables using the sample variance s^2 (equation 2.3), a statistic that is distributed independently of $\bar{X}$. The t-statistic (section 3.1) formed from such a random sample is itself a random variable

$$t = \frac{\sqrt{n}(\bar{X} - \mu)}{s}.$$

If μ is correct then t has a *Student's t-distribution* with $\nu = n - 1$ degrees of freedom, which has the density function

$$p(x|\nu) = \frac{\Gamma([\nu + 1]/2)}{\sqrt{\nu\pi}\,\Gamma(\nu/2)} (1 + x^2/\nu)^{-(\nu+1)/2}. \tag{5.35}$$

If X is a continuous random variable with this density function, we write

$$X \sim t_\nu.$$

The distribution is symmetric around zero, and so has expectation $\mathrm{E}[X] = 0$. For $\nu > 2$ the variance is $\mathrm{V}[X] = \nu/(\nu - 2)$. The t-distribution resembles a normal distribution with slightly 'fatter' tails (see Figure 5.7). As $\nu \to \infty$ the distribution approaches the standard normal $N(0, 1)$.

5.3.4 Uniform distribution

This is simply the distribution for equally probable values over some finite range $[a, b]$. See Figure 5.8. If X is a continuous random variable following a uniform distribution, then its density is

$$p(x|a, b) = 1/(b - a) \quad \text{for } a \le x \le b \tag{5.36}$$

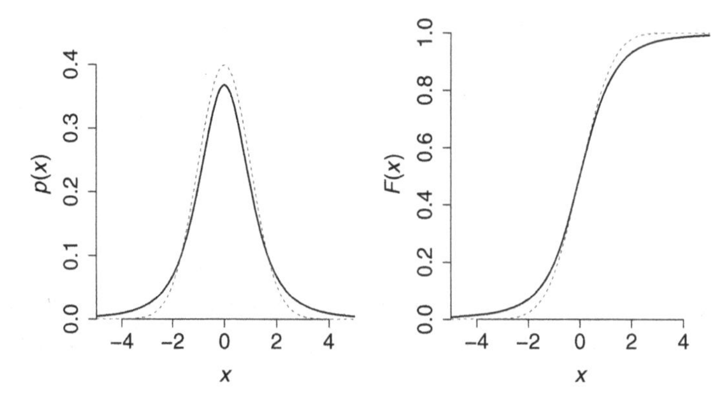

Figure 5.7 Probability density function (pdf; left) and cumulative density functions (cmf; right) for Student's t-distributions with $\nu = 3$ (thick black) compared with the standard normal $N(0, 1)$ (dashed grey).

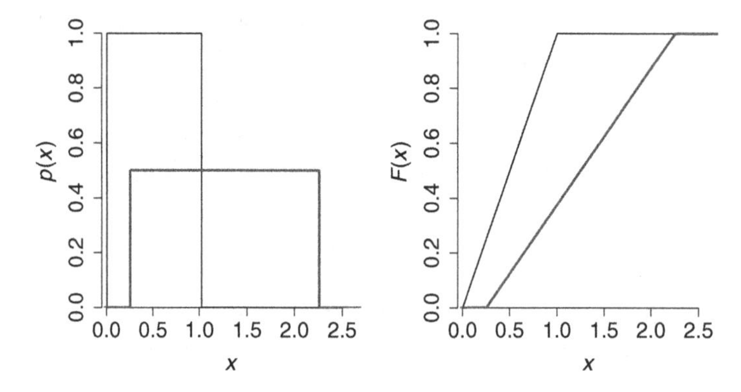

Figure 5.8 Left: uniform pdfs for $U(0, 1)$ (black) and for $U(0.25, 2.25)$ (thick grey). Right: the corresponding cdfs.

and we would write

$$X \sim U(a, b).$$

By applying the definitions of expectation and variance, it is simple to show that $\mathrm{E}[X] = (b - a)/2$ and $\mathrm{V}[X] = (b - a)^2/12$.

5.4 Change of variables

Suppose we know the density function $p_X(x)$ for some random variable X, but we wish to know the density for some transformation of this, i.e. if $Y = f(X)$ we want to know $p_Y(y)$. We need to know how to find the probability function after a change of variables.

If X is discrete, then the transformation is trivial so long as the mapping from X to Y is one-to-one. We simply substitute the values $Y = f(X)$. For example, if X is the score from a single roll of a six-sided die, and $Y = 1/X$, then we can simply make the substitution for each value. The probabilities for the discrete events remain the same: $\Pr(X = x)$ is the same as $\Pr(Y = f(x))$.

For example, imagine rolling two dice and adding the scores. We can model the total score using the random variable X, where $\Pr(X = 2) = 1/36$ and $\Pr(X = 7) = 6/36$, and so on. Now, if we make the change of variables $Y = 1/X$ we find that $\Pr(X = 2) = \Pr(Y = 1/2) = 1/36$ and so on. The probabilities are unchanged; we have essentially just re-labelled the possible values of the random variable.

The situation for continuous variables requires more thought. Here we must find the way that probability density functions transform upon a change of variables. The simplest case is where X is a continuous random variables with density $p_X(x)$, and $Y = f(X)$ is an increasing or decreasing function. This means there is a one-to-one correspondence between the possible values of each variable, and we can write the inverse transformation as $X = g(Y)$. In this case the basic relation is

$$|p(x)\mathrm{d}x| = |p(y)\mathrm{d}y| \tag{5.37}$$

but we can build this more carefully as follows. The event that Y is found in the interval (a, b) is the same as the event that X is found in the interval $(g(a), g(b))$. And so

$$\begin{aligned} \Pr(a < Y < b) &= \Pr(g(a) < X < g(b)) \\ \int_a^b p_Y(y)\mathrm{d}y &= \left| \int_{g(a)}^{g(b)} p_X(x)\mathrm{d}x \right| \\ &= \int_a^b p_X(x) \left| \frac{\mathrm{d}x}{\mathrm{d}y} \right| \mathrm{d}y. \end{aligned} \tag{5.38}$$

We need to use the absolute value to ensure the integral is positive in the case that $f(X)$ is a decreasing function. If we pull out the integrands, we find

$$p_Y(y) = p_X(x) \left| \frac{\mathrm{d}x}{\mathrm{d}y} \right| = p_X(x) \left| \frac{\mathrm{d}f(x)}{\mathrm{d}x} \right|^{-1} = \frac{p_X(g(y))}{|f'(g(y))|} \tag{5.39}$$

using $f'(x) = \mathrm{d}y/\mathrm{d}x$. In the multi-dimensional case, $|\mathrm{d}x/\mathrm{d}y|$ is replaced by the Jacobian determinant of the transformation $\mathbf{Y} = f(\mathbf{X})$. (We assume the relevant derivatives exist.)

If the transformation is not one to one, and there are many possible X values that transform to the same Y value, then we must sum over the corresponding patches. An example should help make this clear. If $Y = |X|$, then both $X = -2$ and $X = 2$

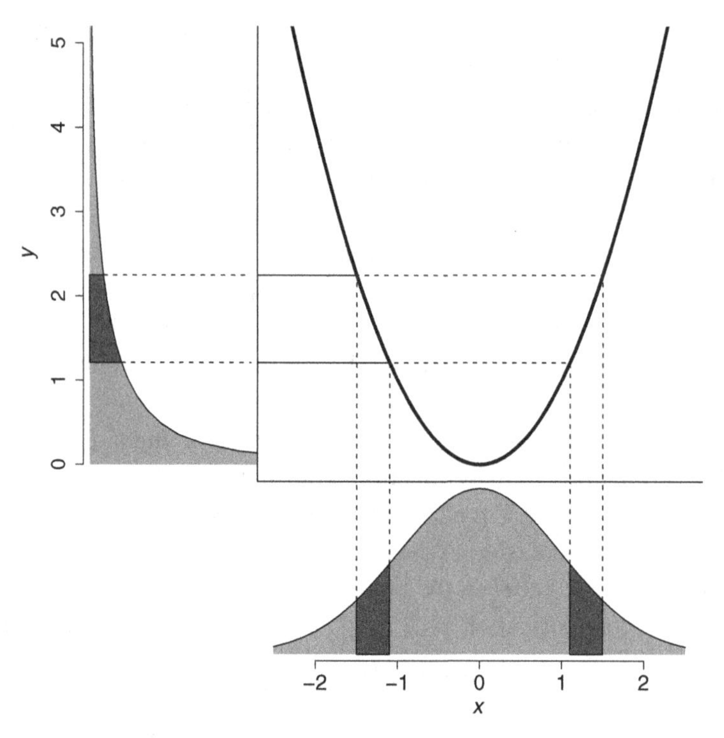

Figure 5.9 Illustration of the transformation of variables. The upper-right panel shows the relation $y = x^2$ (thick curve). The lower panel shows the (standard) normal distribution of random variable X. The left panel shows the distribution after the transformation $Y = X^2$. The shaded areas show how an interval on Y ($a < Y < b$) maps onto intervals on X.

map onto the same Y value. The cumulative distribution of Y is

$$\begin{aligned} F_Y(y) &= \Pr(Y \le y) \\ &= \Pr(|X| \le y) \\ &= \Pr(-y \le X \le y) \\ &= F_X(y) - F_X(-y) \end{aligned} \tag{5.40}$$

where the last line is an application of equation 4.27. Now, from equation 4.29 we know that the derivative of the cumulative distribution function gives the density, so we differentiate $F_Y(y)$ to find

$$p_Y(y) = \begin{cases} p_X(y) + p_X(-y) & \text{for } y > 0 \\ 0 & \text{otherwise.} \end{cases} \tag{5.41}$$

We have summed the density for each value of X that transforms to a given value for Y.

Now we can use this to examine another important transformation. If $X \sim N(0, 1)$, then what is the distribution of $Z = X^2$? We can do this by first making the transformation $Y = |X|$, as we did above, then $Z = Y^2$. The first transformation deals with the fact that the mapping from X is not one to one. Now we must find the distribution of Z from equation 5.41 and the normal density function equation 5.30.

$$p(y) = p(x) + p(-x) = 2N(0, 1) = \frac{2}{\sqrt{2\pi}} e^{-y^2/2} \tag{5.42}$$

for $y > 0$. In order to use the change-of-variable equation (5.38), we must find the relevant derivative – $dy/dz = z^{-1/2}/2$ – and then substitute $z = y^2$ into equation 5.42 to get

$$p(z) = \frac{2}{\sqrt{2\pi}} e^{-z/2} \left| \frac{z^{-1/2}}{2} \right| = \frac{z^{-1/2} e^{-z/2}}{\sqrt{2\pi}} \tag{5.43}$$

for $z > 0$. This is exactly the same as the density function for the χ^2 distribution with $\nu = 1$ degrees of freedom (equation 5.33), as we would expect from our definition of a χ^2 variable (above). This particular transformation is illustrated in Figure 5.9.

R.Box 5.11

Testing the transformation relation

We can test the transformation relation by making a large sample of random data from some distribution (e.g. standard normal), transforming the results, and looking at the distribution of the transformed data. First, we shall produce a sample of normal data and plot its histogram

```
x <- rnorm(1E4)
h <- hist(x, breaks=100, col="grey", prob=TRUE)
pdf.norm <- dnorm(h$mids)
lines(h$mids, pdf.norm, col="red", lwd=4)
```

The last two lines compute and plot the corresponding values for the (standard normal) density curve. Now, if we make the transformation $z = x^2$, we should (following the above analysis) find the data z follows a χ^2 (with $\nu = 1$ degrees of freedom) distribution.

```
z <- x^2
h <- hist(z, breaks=100, col="grey", prob=TRUE)
pdf.chisq <- dchisq(h$mids, df=1)
lines(h$mids, pdf.chisq, col="red", lwd=4)
```

5.5 Approximate variance relations (or the propagation of errors)

The method of transformation (above) works when we know the distribution of X and wish to find the distribution of $Z = f(X)$. There are some cases where we do not know the distribution of X fully, but we can use its first two moments (mean and variance) to approximate the mean and variance of Z. We can use this same method for functions of more than one variable, such as $Z = f(X, Y)$. In the special case of a linear function of random variables, we can compute the variance exactly (section 5.1.4), but the following approximate method will work for more general functions.

Suppose we have random variables $\mathbf{X} = \{X_1, X_2, \ldots, X_n\}$ and we know (or have estimates of) their means and variances ($\mathrm{E}[\mathbf{X}] = \boldsymbol{\mu} = \{\mu_1, \mu_2, \ldots, \mu_n\}$ and $\mathrm{V}[\mathbf{X}] = \boldsymbol{\sigma}^2 = \{\sigma_1^2, \sigma_2^2, \ldots, \sigma_n^2\}$). More generally we might have a covariance matrix for all the variables Σ_{ij} (see equation 5.12). From this information, we wish to estimate the mean and variance of the function $Z = f(\mathbf{X})$.

The first step is to write down the Taylor series expansion of the function f about the point $\boldsymbol{\mu}$

$$Z = f(\mathbf{X}) = f(\boldsymbol{\mu}) + \sum_{i=1}^{n} \left.\frac{\partial f}{\partial X_i}\right|_{\mathbf{X}=\boldsymbol{\mu}} (X_i - \mu_i) + \cdots. \tag{5.44}$$

If we neglect the higher-order terms, and use only the first-order approximation, we can see that $\mathrm{E}[Z] \approx f(\boldsymbol{\mu})$ (the expectation of the second term in the Taylor series is zero since $\mathrm{E}[X_i - \mu_i] = 0$). In order to find the corresponding approximate formula for the variance, we note that $\mathrm{V}[f(X)] = \mathrm{E}[f(X)^2] - \mathrm{E}[f(X)]^2$ (equation 5.7) and we already have an approximation for the second term ($\approx f(\boldsymbol{\mu})^2$). It remains to find $\mathrm{E}[f(X)^2]$, which we can do by taking the first-order Taylor series above and squaring.

$$f(\mathbf{X})^2 \approx f(\boldsymbol{\mu})^2 + 2f(\boldsymbol{\mu}) \sum_{i=1}^{n} \left.\frac{\partial f}{\partial X_i}\right|_{\mathbf{X}=\boldsymbol{\mu}} (X_i - \mu_i)$$
$$+ \left[\sum_{i=1}^{n} \left.\frac{\partial f}{\partial X_i}\right|_{\mathbf{X}=\boldsymbol{\mu}} (X_i - \mu_i)\right] \left[\sum_{j=1}^{n} \left.\frac{\partial f}{\partial X_j}\right|_{\mathbf{X}=\boldsymbol{\mu}} (X_j - \mu_j)\right]. \tag{5.45}$$

When we take the expectation the second term in the sum becomes zero:

$$\mathrm{E}[f(\mathbf{X})^2] \approx f(\boldsymbol{\mu})^2 + \sum_{i=1}^{n}\sum_{j=1}^{n} \left.\frac{\partial f}{\partial X_i}\frac{\partial f}{\partial X_j}\right|_{\mathbf{X}=\boldsymbol{\mu}} \Sigma_{ij} \tag{5.46}$$

where Σ_{ij} is the (i, j)th element of the covariance matrix, i.e. σ_{ij}^2 (see equation 5.12). Using this we can approximate the variance of Z as

$$V[Z] = V[f(X)] = \mathrm{E}[f(X)^2] - \mathrm{E}[f(X)]^2 \approx \sum_{i=1}^{n}\sum_{j=1}^{n} \left.\frac{\partial f}{\partial X_i}\frac{\partial f}{\partial X_j}\right|_{\mathbf{X}=\boldsymbol{\mu}} \Sigma_{ij} \quad (5.47)$$

and in the special case where the variables X_i are all uncorrelated ($\Sigma_{ij} = 0$ for $i \neq j$) we get

$$V[Z] = \sigma_z^2 \approx \sum_{i=1}^{n} \left(\frac{\partial f}{\partial X_i}\right)^2_{\mathbf{X}=\boldsymbol{\mu}} \sigma_i^2 \quad (5.48)$$

where $\Sigma_{ii} = \sigma_i^2$ are the variances of the variables X_i (the diagonal elements of the covariance matrix).

These are the formulae used to *propagate uncertainty* (a subject that bedevils students in undergraduate laboratory classes). They are useful when we have estimates of some variables and associated standard deviations (the 'uncertainties'), but wish to estimate the standard deviation of a function of these. A few of the commonly used special cases are

$$Z = X + Y \quad \text{or} \quad Z = X - Y \Rightarrow \sigma_z^2 = \sigma_x^2 + \sigma_y^2$$

$$Z = XY \Rightarrow \frac{\sigma_z^2}{Z^2} = \frac{\sigma_x^2}{X^2} + \frac{\sigma_y^2}{Y^2}$$

$$Z = X/Y \Rightarrow \frac{\sigma_z^2}{Z^2} = \frac{\sigma_x^2}{X^2} + \frac{\sigma_y^2}{Y^2}$$

assuming X and Y are uncorrelated.

A simple example: we wish to estimate the resistance of a resistor using Ohm's law, $R = V/I$, and estimates and uncertainties of V and I. If V and I are estimated independently (specifically, the measurements are not correlated) as $\hat{V}$ and $\hat{I}$, perhaps from the sample mean of many individual measurements, and we can estimate their standard deviations, σ_V and σ_I, then we can estimate R and its standard deviation using equation 5.48:

$$\hat{R} = \frac{\hat{V}}{\hat{I}} \quad \text{and} \quad \frac{\sigma_R^2}{\hat{R}^2} = \frac{\sigma_V^2}{\hat{V}^2} + \frac{\sigma_I^2}{\hat{I}^2} \Rightarrow \sigma_R = |\hat{R}|\sqrt{\frac{\sigma_V^2}{\hat{V}^2} + \frac{\sigma_I^2}{\hat{I}^2}}. \quad (5.49)$$

Equation 5.47 is only valid to the extent that the first-order Taylor series approximation (equation 5.44) is valid, i.e. that the function $f(\mathbf{X})$ is close to linear in the region of $\boldsymbol{\mu}$ (i.e. within $\pm\boldsymbol{\sigma}$). Equation 5.48 makes the further assumption that the variables are uncorrelated. Where possible it is usually better to use the transformation method (section 5.4), the exact variance relations (e.g. section 5.1.4), or the

Monte Carlo method (Chapter 8), to estimate the statistical properties of a function of estimated variables.

5.6 Chapter summary

- For a random variable X, the expectation of a function $f(X)$ is defined as

$$\mathrm{E}[f(X)] = \int_{-\infty}^{+\infty} f(x)p(x)\mathrm{d}x.$$

The mean μ and variance σ^2 are the expectation of X and $(X - \mu)^2$, respectively:

$$\mu_x = \mathrm{E}[X] = \int_{-\infty}^{+\infty} xp(x)\mathrm{d}x$$

$$\sigma_x^2 = \mathrm{E}[(X - \mu_x)^2] = \int_{-\infty}^{+\infty} (X - \mu_x)^2 p(x)\mathrm{d}x.$$

- The covariance σ_{xy} between X and Y, and correlation coefficient ρ,

$$\sigma_{xy} = \mathrm{E}[(X - \mu_x)(Y - \mu_y)] = \mathrm{E}[XY] - \mu_x\mu_y$$

$$\rho(X, Y) = \sigma_{xy}/\sigma_x\sigma_y.$$

- Two variables X and Y are independent if and only if $p(x, y) = p(x)p(y)$, implying $\sigma_{xy} = 0$.
- Binomial distribution for the probability of x successes in n success/failure trials, where the probability of a success in each trial is θ ($x \leq n$):

$$p(x|n, \theta) = \binom{n}{x}\theta^x(1 - \theta)^{n-x}.$$

- Poisson distribution for integer number of events x when the events occur randomly in time and the expected rate (number of events per unit time) is λ:

$$p(x|\lambda) = \frac{\lambda^x \mathrm{e}^{-\lambda}}{x!}.$$

- Central limit theorem: under quite general conditions the distribution of the sum of many independent random variables approaches the normal.
- The normal (or Gaussian) distribution with mean μ and variance σ^2:

$$p(x|\mu, \sigma^2) = \frac{1}{\sigma\sqrt{2\pi}}\mathrm{e}^{-(x-\mu)^2/(2\sigma^2)};$$

the standard normal distribution has $\mu = 0$ and $\sigma^2 = 1$.

- The chi-square distribution with ν degrees of freedom

$$p(x|\nu) = \frac{(1/2)^{\nu/2}}{\Gamma(\nu/2)} x^{\nu/2-1} e^{-x/2}$$

has mean ν and variance 2ν. This is the distribution of the sum of ν variables, each of which is the square of a (standard) normal variable.
- Student's t-distribution with ν degrees of freedom

$$p(x|\nu) = \frac{\Gamma([\nu+1]/2)}{\sqrt{\nu\pi}\,\Gamma(\nu/2)} (1 + x^2/\nu)^{-(\nu+1)/2}$$

has mean 0 and variance $\nu/(\nu-2)$ for $\nu > 2$.
- The uniform distribution between a and b

$$p(x|a,b) = 1/(b-a)$$

has mean $(a+b)/2$ and variance $(b-a)^2/12$.
- The density function for the transformed variable $Y = f(X)$ can be obtained by applying

$$p_Y(y) = p_X(x) \left| \frac{dx}{dy} \right|$$

where $p_X(x)$ is the density function of X.
- The formula for the 'propagation of uncertainty' through the function $Z = f(X, Y)$:

$$\sigma_z^2 \approx \left(\frac{\partial f}{\partial X} \right)^2 \sigma_x^2 + \left(\frac{\partial f}{\partial Y} \right)^2 \sigma_y^2$$

under the assumption that X and Y are uncorrelated, and the function $f(X, Y)$ is approximately linear in the region of interest.

6
Estimation and maximum likelihood

> The sciences do not try to explain, they hardly even try to interpret, they mainly make models. By a model is meant a mathematical construct which, with the addition of certain verbal interpretations, describes observed phenomena. The justification of such a mathematical construct is solely and precisely that it is expected to work ....
>
> John Von Neumann,
> *Method In The Physical Sciences*, 1955

A great deal of data analysis in science involves fitting models to data. Our job in this chapter is to explain a powerful method for estimating the parameters of models by *fitting to data*. This is an extension of the linear regression we looked at in section 3.3. Our method of choice is called *maximum likelihood estimation*.

6.1 Models

Before we move to the main topic of model fitting, let's think carefully about what we need from a model. In part, the model should encapsulate the physics of the situation under scrutiny. We usually wish our models to be in the form of mathematical relationships between explanatory variables and one or more response variables. In the case of Reynolds' pipe experiment (section B.3), we have

$$v = \left(\frac{R^2}{8\eta}\right)\frac{\Delta P}{\Delta L}, \tag{6.1}$$

which relates the response variable v (velocity of the flow) to an explanatory variable (in this case $\Delta P/\Delta L$, the pressure gradient). The other terms in the relationship are either constants (e.g. mathematical or physical constants), or other factors that are determined by the experimental set-up but are not themselves considered variable (such as R, the radius of the pipe, and η, the viscosity of water, which depends on temperature).

For the purposes of statistical data analysis, we need to consider another aspect of the model. A model as described above would be valid for making predictions about the expected value of the response variable, but tells us nothing else about the distribution of the experimental data. We therefore need to have a *statistical* model to complement the *physics* model, and that encapsulates what we know about the data collection process. Often in laboratory work this will comprise estimates of the standard deviation of the measurements ('error bars') and the assumption that the data are normally distributed. For example, we may know (from prior experimental testing) that the distribution of the v measurements is approximately normal with $\sigma = 0.005 \text{ m s}^{-1}$. The statistical model captures mathematically what we know about the randomness of the data collection process. The randomness may be due to 'measurement noise' or it may be intrinsic to the physical nature of the experimental subject (e.g. random radioactive decays).

If the model has no unknown terms, then it specifies a *simple* hypothesis. From the model alone we can predict the distribution of the data (response variable) as a function of the explanatory variables. In this case the task of statistical data analysis is testing, or checking the consistency between data and model, which is covered in the next chapter. If, on the other hand, there are terms in the model whose numerical values are not known to high precision, then the model is said to specify a *composite* hypothesis, and the terms with unknown values are the model's *free parameters*. In the case of Reynolds' model, we may know R accurately but not η, in which case η is a free parameter, whose value we can estimate using the data. It is this business of estimating model parameters given some data and a model that is the subject of this chapter.

6.2 Case study: Rutherford & Geiger data

Let's take another look at data from the Rutherford & Geiger experiment (section B.2). The authors discussed a specific model for the data, namely that there is a constant average rate of polonium scintillations λ (in units of counts per interval) but that the number recorded in any short time interval follows a Poisson distribution. This tells us the probability of observing data x (counts per interval), if the true (expected) rate is λ. This kind of equation, predicting the distribution of the data, is often called the *sampling distribution* of the data. If we knew λ, we could predict the distribution of data x. But we do not know λ; we only have observations of the scintillations per interval recorded by Rutherford & Geiger, and wish to estimate λ, the one free parameter of our model.

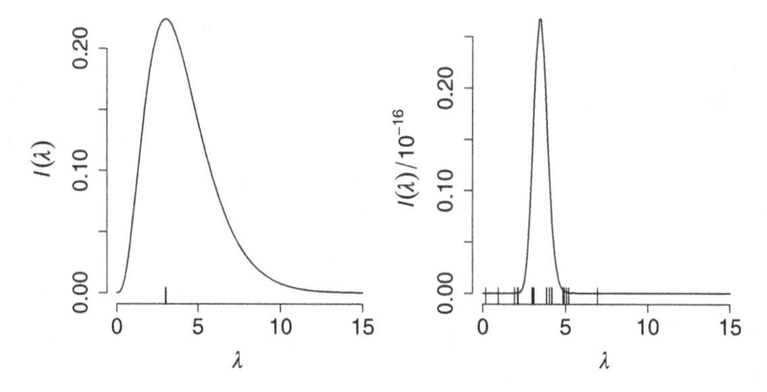

Figure 6.1 The Poisson likelihood $l(\lambda)$ as a function of λ given (left) a single data point $x^{\text{obs}} = 3$, and (right) $n = 20$ data points. The data points are indicated using a 'rug' on the horizontal axis ('jittered' – for plotting purposes only – so points do not overlap). In the $n = 1$ case, the likelihood peaks at $\lambda = 3$, i.e. the same value as the single datum. In the $n = 20$ case, the likelihood function has become narrower and the peak has moved to 3.45 (this likelihood function has also been rescaled by a factor of 10^{16}).

6.2.1 One data point

Let's run the experiment for just one interval and collect one data point: $x^{\text{obs}} = 3$ was the first in the sample published by Rutherford & Geiger. We now have some information on the decay rate.

We know that for any given λ there will be some values of x that are more probable that others. An intuitively reasonable approach is to choose the value of λ that would have made our observed value, x^{obs}, highly probable. In order to do this, we use the Poisson function (equation 5.29) but treat it as a function of the parameter λ, with $x = x^{\text{obs}}$ fixed. When we use a distribution function as a function of its parameters (with x fixed) we call it a *likelihood* function.[1] The Poisson likelihood, $l(\lambda)$, with $x^{\text{obs}} = 3$ is

$$l(\lambda) = \frac{\lambda^3 e^{-\lambda}}{3!}. \tag{6.2}$$

Our estimate of the decay rate, $\hat{\lambda}$, is the value of λ that maximises this function, the *mode* of the likelihood function. Such an estimate is called a *maximum likelihood estimate* (MLE). The caret ('hat' symbol) is used to distinguish our estimate of the parameter from the parameter itself. Figure 6.1 (left) shows

[1] The words 'probability' and 'likelihood' are synonymous in everyday usage, but they have different technical meanings. Once the data have been taken we should not really talk about the probability distribution of the data, since the data values are now known, no longer random. When we use a probability formula $p(\cdot|\cdot)$ as a function of the second (conditional) argument, not the first argument, we call it a likelihood function. The formulae for probability and likelihood are the same, but they are functions of different things: probability of the data (given the parameters), likelihood of the parameters (given the data).

the likelihood function from a single observation $x^{\text{obs}} = 3$. Clearly the mode is around 3.

The location of the maximum of the likelihood function can be found by the usual route: find the value of λ at which the derivative is zero,[2] which in this case gives

$$\frac{\partial l(\lambda)}{\partial \lambda} = (x\lambda^{x-1} - \lambda^x)\frac{\mathrm{e}^{-\lambda}}{x!} = 0$$

which leads immediately to the solution. Given some data x^{obs} the MLE is $\hat{\lambda} = x^{\text{obs}}$. Of course, this was obvious: with one observation of the decay rate $x^{\text{obs}} = 3$ the maximum likelihood estimate (MLE) of the decay rate is also $\hat{\lambda} = 3$. Another value of λ, one far away from $\hat{\lambda}$, would make the data less likely to have happened.

6.2.2 Many data points

Now suppose we go back to our experiment and take more data. We now have many x_i $(i = 1, 2, \ldots, n)$ but still only one λ to estimate. We can arrive at the MLE of λ by the same argument as before. The sampling distribution for each datum x_i is $p(x_i|\lambda)$, as before. The sampling distribution for two independent observations x_i and x_j is, by the multiplication rule, simply the product of the two separate sampling distributions (for independent variables, equation 4.12). By extension, the sampling distribution for obtaining a dataset $\mathbf{x} = \{x_1, x_2, \ldots, x_n\}$ (i.e. observing x_1 and x_2 and so on ...) is given by the product of the distribution for each datum x_i.

$$\begin{aligned} p(\mathbf{x}|\lambda) &= p(x_1, x_2, \ldots, x_n|\lambda) \\ &= p(x_1|\lambda) \times p(x_2|\lambda) \times \cdots \times p(x_n|\lambda) \\ &= \prod_{i=1}^{n} p(x_i|\lambda) = \prod_{i=1}^{n} \frac{\lambda^{x_i}\mathrm{e}^{-\lambda}}{x_i!}. \end{aligned} \tag{6.3}$$

The likelihood function $l(\lambda)$ is simply this formula considered as a function of its parameter λ for fixed data $\mathbf{x} = \mathbf{x}^{\text{obs}}$. Once we insert the data, we can find the mode of $l(\lambda)$ to obtain the MLE, i.e., the value for the parameter that makes the whole dataset most likely. Figure 6.1 (right) shows the likelihood function calculated from the first 20 observations of the Rutherford and Geiger experiment. Clearly this is narrower than before, and peaks at 3.45.

Finding the mode of $l(\lambda)$ can be a challenging task, but there is a trick that is useful in many situations: instead of finding the maximum of $l(\lambda)$, we find

[2] We should also check it is a maximum by confirming that $\partial^2 l/\partial\lambda^2 < 0$. If there are several solutions, we select the one that gives the maximum $l(\lambda)$.

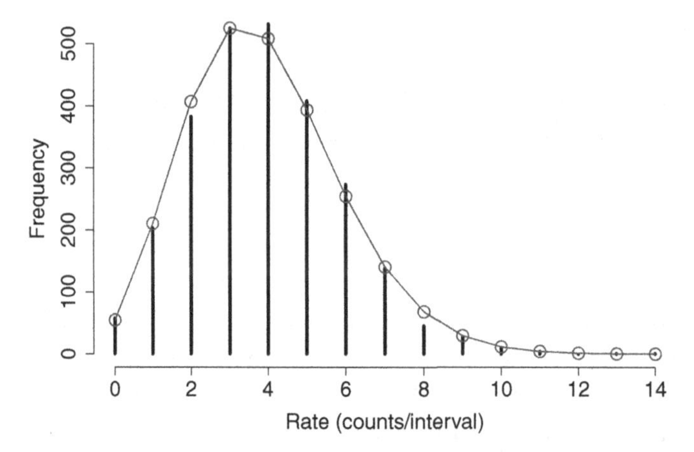

Figure 6.2 The Rutherford & Geiger data (vertical bars) compared with a (scaled) Poisson distribution with rate $\lambda = 3.87$ (circles joined by lines).

the maximum of its logarithm.[3] Taking the logarithms makes the maths easier but changes nothing important about the problem. Finding the mode of $L(\lambda) = \log l(\lambda)$ is the same as finding the mode of $l(\lambda)$. (We'll use the natural logarithm, although other bases would work fine.)

$$\begin{aligned}
L(\lambda) &= \log l(\lambda) = \log[p(\mathbf{x}|\lambda)] \\
&= \log \prod_{i=1}^{n} \frac{\lambda^{x_i} e^{-\lambda}}{x_i!} = \sum_{i=1}^{n} \log \frac{\lambda^{x_i} e^{-\lambda}}{x_i!} \\
&= \sum_{i=1}^{n} \{x_i \log(\lambda) - \lambda - \log(x_i!)\} \\
&= \left(\sum_{i=1}^{n} x_i \right) \log(\lambda) - n\lambda - \sum_{i=1}^{n} \log(x_i!).
\end{aligned} \tag{6.4}$$

The sum is over the n observations of the rate x_i. The sum is easier to differentiate than the product we started with. Notice the last term in the above expression is a constant (it has no dependence on the parameter λ). Now we need to find the value of λ for which the derivative $\partial L / \partial \lambda$ is zero; call this $\hat{\lambda}$:

$$\left. \frac{\partial L(\lambda)}{\partial \lambda} \right|_{\lambda = \hat{\lambda}} = \left(\sum_{i=1}^{n} x_i \right) \frac{1}{\hat{\lambda}} - n = 0 \quad \Rightarrow \quad \hat{\lambda} = \frac{1}{n} \sum_{i=1}^{n} x_i^{\text{obs}} = \bar{x}. \tag{6.5}$$

[3] Here we use log to indicate the natural logarithm (base e). The corresponding R command is simply `log()`. For base 10 use `log10()`.

In other words, the maximum likelihood estimate of the decay rate is just the sample mean of the observed rates. For the full Rutherford–Geiger data ($n = 2608$), this gives a maximum likelihood estimate (MLE) for the scintillation rate of $\hat{\lambda} = 3.87$ counts/interval.

Note that this is the MLE of the parameter λ for the Poisson model, i.e. we are assuming the Poisson model is right (or, at least, useful). At this stage, we should check whether this model looks sensible by plotting the data and the model Poisson distribution for rate $\lambda = 3.87$ (see Figure 6.2). In this case, the model does seem to match the data well.

R.Box 6.1

A model comparison for the Rutherford–Geiger data

Section B.2 describes data from the Rutherford and Geiger experiment, and shows how to load and plot the data, and compute the mean count rate of 3.87 counts/interval. The mean count rate is the MLE for the one parameter of the (Poisson) model for the data. The following lines show how we can define a simple function to compute the Poisson model:

```
model.pois <- function(x, lambda, scale) {
  mod.y <- dpois(x, lambda=lambda) * scale
  return(mod.y)
}
```

Notice that we used the `dpois()` function to compute the Poisson pmf, but then multiplied by the number of intervals. This is because the Poisson function sums to unity, but Rutherford and Geiger recorded 2608 intervals, so the histogram of the data sums to 2608. Then we can use the newly defined function and overlay its output on a plot of the data (the first plot from R.box 2.2).

```
mod.y <- model.pois(rate, lambda=mean.rate, scale=n.obs)
lines(rate, mod.y, type="o", col="red", lwd=3)
```

The result is shown in Figure 6.2. The model appears to match the data well.

6.3 Maximum likelihood estimation

R.Box 6.2

Fitting Reynolds' data I: preparing the data R

Over the next few R.boxes we shall build a complete script to find the maximum likelihood estimates (MLEs) for the parameters of a simple model given some data. The commands can all be entered directly at the command line, but it is good practice to write them into a script – this can be checked, changed, re-run and extended as needed (see section A.6). First we load Reynolds' data and extract the first eight data points

```
# load the data
  dat <- read.table("fluid.txt", header=TRUE)
  x <- dat$dP[1:8]
  y <- dat$v[1:8]
  dy <- 6.3e-3
```

The data comprise three variables we shall call x, y and dy: x is the explanatory variable, y is the response variable and dy is the uncertainty associated with y, which is entered 'by hand' in this case.

The procedure used above – estimating the parameter of a model by finding the value that maximises the likelihood – can in principle be used for a very wide range of data and models. We first write down the likelihood function for the parameters, given the data, and then find the parameter values that maximise this function. The likelihood function is the probability distribution of the data (the sampling distribution), which we can build up in two parts, which we can call the physical model and the probability model. The physical model predicts the expectation of the data, based on understanding of the physics; the probability model predicts the sampling distribution of the data about the expectation, based on understanding of the data collection process.

Consider an experiment from which we obtain measurements of a response variable y_i at different values of a response variable x_i for $i = 1, 2, \ldots, n$. For compactness we write these as vectors, $\mathbf{y}$ and $\mathbf{x}$. From consideration of previous experiments and theory, we have a theoretical model that predicts the expectation of y as a function of x: $\mathrm{E}[y] = f(x)$. But this model is not completely specified, that is, it contains some parameters with unknown values. We shall call the parameters $\boldsymbol{\theta} = \{\theta_1, \theta_2, \ldots, \theta_M\}$ and write the model as $f(x, \boldsymbol{\theta})$. The shape of the distribution about the expected value is determined by the sampling distribution for the data, which we can write as $p(\mathbf{y}|\boldsymbol{\theta})$. The distributions covered in Chapter 5 are those most commonly encountered in physics experiments. This is enough to construct the likelihood function.

$$l(\boldsymbol{\theta}) = p(y_1, \ldots, y_n|\boldsymbol{\theta}) = p(y_1|\boldsymbol{\theta}) \times \cdots \times p(y_n|\boldsymbol{\theta}) = \prod_{i=1}^{n} p(y_i|\boldsymbol{\theta}). \tag{6.6}$$

This holds so long as the y_i are independent of one another (multiplication rule; equation 4.12); otherwise, we need to account for the covariances between the

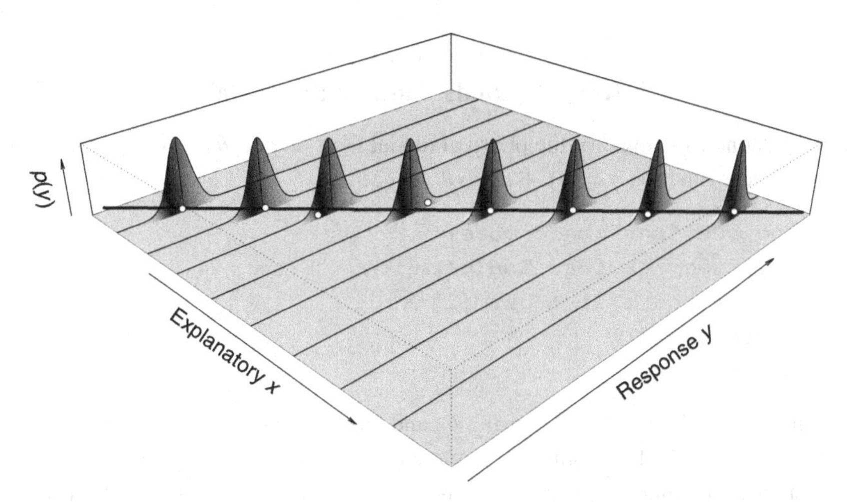

Figure 6.3 Illustration of the idea of maximum likelihood fitting. The observed data are indicated with white circles. The maximum likelihood model, which predicts the expectation of the response variable, $\mathrm{E}[y] = f(x, \boldsymbol{\theta})$, is shown as the thick solid line. Thin black lines show the sampling distributions around the model $p(y_i|\boldsymbol{\theta})$, which in this case are all normal.

response data points. The MLEs for the parameters $\boldsymbol{\theta}$ are those that maximise $l(\boldsymbol{\theta})$ given observational data $\mathbf{y} = \mathbf{y}^{\mathrm{obs}}$.

As mentioned above it is often easier to use the 'log likelihood' function, because the product becomes a sum

$$L(\boldsymbol{\theta}) = \log l(\boldsymbol{\theta}) = \log \prod_{i=1}^{n} p(y_i|\boldsymbol{\theta}) = \sum_{i=1}^{n} \log p(y_i|\boldsymbol{\theta}). \tag{6.7}$$

We wish to find the values of θ_j that maximise this. In simple cases, such as with linear models and normal sampling distributions, the analytical solutions are known, but in general the MLEs must be found using numerical methods. It is worth noticing that $\min L$ (or $\min l$) are statistics, that is, they are single numbers calculated from random data.

The principles underlying maximum likelihood fitting are illustrated in Figure 6.3. We have some data for a response variable y_i taken as a function of an explanatory variable x_i; the data are plotted as white circles in the x–y plane. We then have a physical model that gives the expected y values, $\mathrm{E}[y] = f(x, \boldsymbol{\theta})$, indicated by the solid curve passing through the x–y plane. At each x_i there is a sampling distribution for the data to be distributed about the expected values, $p(y_i|\boldsymbol{\theta})$; these are curves in the y–z plane. The likelihood for the whole dataset is the product of the individual likelihoods $p(y_i = y_i^{\mathrm{obs}}|\boldsymbol{\theta})$. The maximum likelihood estimated parameters $\hat{\boldsymbol{\theta}}$ represent the model that gives the highest overall likelihood.

R.Box 6.3

Fitting Reynolds' data II: the model

We must define functions for the physical model $E[y] = f(x, \boldsymbol{\theta})$, which in this case is a simple linear function $y = bx + a$ (so $\boldsymbol{\theta} = \{a, b\}$).

```
# define the "physical" model
  model.linear <- function(parm, x) {
    y.mod <- parm[1] + parm[2]*x
    return(y.mod)
  }
```

The function is called `model.linear()` and takes two arguments as input, `parm`, and x value(s) at which to compute the model. The vector `parm` contains values for each model parameter: `parm[2]` is the slope, `parm[1]` is the intercept. The function returns as output a value for $E[y] = f(x, a, b)$. Notice that `parm[1]` and `parm[2]` are scalars (single numbers), but if `x` is a vector (a list of numbers) then `y.mod` will be a vector of the same size.

R.Box 6.4

Fitting Reynolds' data III: the minus log likelihood

We need to define the likelihood function we wish to maximise. As discussed elsewhere, the logarithm of the likelihood is a useful function to work with. In fact we shall use the minus log likelihood, and minimise this function. Of course, the position of the minimum of the minus log likelihood is the same as the position of the maximum of the likelihood.

```
# define the fit statistic
  LogLikelihood <- function(parm, x, y, dy, model) {
    y.mod <- model(parm, x)
    l <- dnorm(y, mean=y.mod, sd=dy, log=TRUE)
    mlogl <- -sum(l)
    return(mlogl)
  }
```

The function `LogLikelihood()` takes five arguments as input: `parm`, a vector of parameters, the data contained in `x`, `y`, the errors `dy`, and lastly the name of the model function. The first line calls the function whose name is stored in `model` to calculate the physical model.

The second line inside the function computes the logarithm of the likelihood for each data point y_i. This is done using the `dnorm()` function with the `log=TRUE` argument to give the logarithm of the normal density curve at point `y` (i.e. the data) for mean `y.mod` and standard deviation `dy`. Notice how this one line computes the likelihood values for each data point, with no need for an explicit loop. The last two lines sum the log likelihoods over all data points, take the negative, and return the result.

Box 6.1
Scores

The partial derivatives of the log likelihood function $L(\boldsymbol{\theta})$ with respect to each parameter are collectively known as the *score function* $U(\theta)$

$$U(\boldsymbol{\theta}) = \left(\frac{\partial L(\boldsymbol{\theta})}{\partial \theta_1}, \ldots, \frac{\partial L(\boldsymbol{\theta})}{\partial \theta_M} \right) \tag{6.8}$$

which (you may recall from vector calculus class) can be written with the *grad* differential operator: $U(\boldsymbol{\theta}) = \nabla L$. The maximum likelihood solution $\hat{\boldsymbol{\theta}}$ corresponds to the parameter values that give a zero score

$$U(\hat{\boldsymbol{\theta}}) = (0, \ldots, 0) = \mathbf{0}. \tag{6.9}$$

6.4 Weighted least squares

It is very often the case in the physical sciences that our measurements have a normal distribution. By this we simply mean that the response variable y_i, at each value of the explanatory variable x_i, has the pdf of equation 5.30 with expectation $\mu_i = \mathrm{E}[y_i] = f(x_i, \boldsymbol{\theta})$ and variance σ_i^2. If we have n data points distributed independently, then the response data have the distribution

$$p(\mathbf{y}|\boldsymbol{\mu}, \boldsymbol{\sigma}^2) = \prod_{i=1}^{n} \frac{1}{\sqrt{2\pi\sigma_i^2}} \exp\left\{ -\frac{(y_i - \mu_i)^2}{2\sigma_i^2} \right\} \tag{6.10}$$

where we have used bold characters to indicate vectors, i.e. lists of values (e.g. $\mathbf{y} = \{y_1, \ldots, y_n\}$, and similarly for $\boldsymbol{\mu}, \boldsymbol{\sigma}$).

Once we have specified the physics model $\boldsymbol{\mu} = \mathrm{E}[\mathbf{y}] = f(\mathbf{x}, \boldsymbol{\theta})$, which depends on the unknown parameters $\boldsymbol{\theta} = \{\theta_1, \ldots, \theta_M\}$ we can write down the log likelihood function (see equation 6.7).

$$L(\boldsymbol{\theta}) = \log p(\mathbf{y}|\boldsymbol{\mu}, \boldsymbol{\sigma}^2) = -\frac{1}{2} \sum_{i=1}^{n} \log\left[2\pi\sigma_i^2\right] - \frac{1}{2} \sum_{i=1}^{n} \frac{(y_i - \mu_i)^2}{\sigma_i^2}. \tag{6.11}$$

Notice the first term of the right side is a constant, once the data are fixed, because it does not depend on the parameters $\boldsymbol{\theta}$. If we drop this constant term (since it makes no difference to the maximisation), we can consider a new function

$$X^2(\boldsymbol{\theta}) = -2L(\boldsymbol{\theta}) + const = \sum_{i=1}^{n} \frac{(y_i - \mu_i)^2}{\sigma_i^2}. \tag{6.12}$$

This is the sum of the squared *data – model* variations, weighted by the precisions (i.e. $1/\sigma_i^2$). The *data – model* differences are often called the *residuals* (see section 3.3). If we minimise the function $X^2(\boldsymbol{\theta})$, it is the same as maximising the log likelihood $L(\boldsymbol{\theta})$, or indeed the likelihood $l(\boldsymbol{\theta})$, since

$$l = \exp(L) \propto \exp(-X^2/2). \tag{6.13}$$

The process of finding the parameters that minimise the function $X^2(\boldsymbol{\theta})$ is sometimes known as *weighted least squares*. The reason is that we search for the parameter values that give the smallest values for the weighted sum of the square residuals. When the data are normally distributed this is equivalent to maximising the likelihood. The function $X^2(\boldsymbol{\theta})$ is often known as the *chi square* fit statistic (sometimes written $\chi^2(\boldsymbol{\theta})$) for reasons that will be discussed in Chapter 7.

R.Box 6.5
Fitting using weighted least squares

If, instead of using the `LogLikelihood()` function for direct maximum likelihood, you prefer to use weighted least squares (equation 6.12), then the following code will define the relevant function. The inputs are the same as for the `LogLikelihood()` function (above).

```
# define chi square (least squares) statistic
ChiSq <- function(parm, x, y, dy, model) {
  y.mod <- model(parm, x)
  X <- sum( (y - y.mod)^2 / dy^2)
  return(X)
}
```

R.Box 6.6
Fitting Reynolds' data IV: the fitting

Now we have data, a physical model and a suitable function (the minus log likelihood function) that needs to be minimised. What we need now is a way to minimise it. R has a suite of functions for minimisation. One of the most useful is `optim()` for non-linear minimisation.

```
parm.0 <- c(0.0, 0.1)
result.reyn <- optim(fn=LogLikelihood, par=parm.0,
                     hessian=TRUE, x=x, y=y, dy=dy,
                     model=model.linear)
print(result.reyn)
```

We specify an initial guess for the parameters, in this case $\boldsymbol{\theta} = (a, b)$ and we start with $a = 0$ and $b = 0.1$. Notice that the parameters are listed in the vector `parm.0` in the order they are specified in the definition of the model (see R.box 6.3). If we get crazy results using this starting position, we can try a different guess as the starting position. (The `hessian=TRUE` argument will come in useful later.)

Then we use the `optim()` function to perform the minimisation. The first two arguments to this are the name of the function to be minimised (`LogLikelihood`) and a vector of starting values for the parameters (`parm.0`). The final arguments (`x=x`, etc.) are any additional inputs to the function being minimised (`x`, `y`, `dy`, `model` – see R.box 6.4). The minimisation works by iteratively adjusting the parameters, starting from the initial values, in the direction that causes the function to decrease, until the rate of decrease becomes sufficiently close to zero, indicating we have found a minimum (or there is a problem!).

If you were fitting by weighted least squares, you would minimise the `ChiSq()` function using e.g. `optim(fn=ChiSq, ...)`.

R.Box 6.7
Fitting Reynolds' data V: the fitting results

The result of the minimisation is a list (`result.reyn`) containing several items, including

- `value` – the minimum value found for the minus log likelihood function
- `par` – the values of the parameters at the minimum
- `convergence` – a diagnostic code (0 = success, see `?optim` for more information)
- `hessian` (optional) – a matrix that will be useful for calculating confidence intervals.

For our purposes the first thing to look at is `result.reyn$par`. Assuming the minimisation worked fine (we can check this later) these are our maximum likelihood estimates (MLEs) for the parameters a and b.

Try repeating the minimisation using different starting value – `parm.0`. This can be a useful check that the minimum you have found is the global minimum and not just some local minimum elsewhere in the parameter space.

In cases where the distributions of each data point y_i are normal with the same variance $\sigma_i^2 = \sigma^2$ the function to be minimised simplifies to

$$X^2(\boldsymbol{\theta}) = \frac{1}{\sigma^2} \sum_{i=1}^{n} (y_i - \mu_i(\boldsymbol{\theta}))^2. \tag{6.14}$$

This is often called *ordinary least squares*. This is the same function we minimised in section 3.3, but now we can justify this choice of function by appeal to the fact that minimising this will maximise the likelihood. However, it should be borne in mind that it is only equivalent to maximum likelihood fitting when the sampling distribution of the data is normal (with equal variances in the case of ordinary least squares).

6.4.1 Case study: Reynolds' fluid data

Let's consider a particular example: fitting Reynolds' data for streamline fluid flow, i.e. the first eight data points of Figure 3.5. The physical model is linear, e.g. $\mathrm{E}[y] = bx + a$. Reynolds' published data give little direct information about the shape of the sampling distribution for the velocity measurements, but we can propose the measurements of the response variable follow a simple normal distribution with standard deviation $\sigma = 6.3 \times 10^{-3}$ m s^{-1} for now, and revise this if necessary later. The method, and therefore the result, is equivalent to that discussed in section 3.6.

R.Box 6.8
Fitting Reynolds' data VI: diagnostic plots

It is usually a good idea to visually inspect the match between the data and model. The simplest thing to do is overlay the model on the data.

```
# plot data and model
  y.mod <- model.linear(result.reyn$par, x)
  plot(x, y, pch=16, bty="n")
  segments(x, y-dy, x, y+dy)
  lines(x, y.mod, col="red")
```

The result should be the same as the simple linear regression, Figure 3.5. Notice that in this case we could have plotted the error bars on the model rather than the data. We calculate the model using the MLEs, i.e. evaluate `model.linear()` using the parameter values stored in `result.reyn$par`. Another plot we can examine is a plot of the residuals.

```
# plot residuals
  res <- y - y.mod
  plot(x, res, pch=16, xlab="x", ylim=c(-0.02, 0.02),
       bty="n", ylab="Residual")
  segments(x, res-dy, x, res+dy)
  abline(h=0, lty=2)
```

R.Box 6.9
Mapping out the likelihood using brute force

In the above R.boxes we fitted Reynolds' data with a linear model possessing two parameters, a slope and an intercept. We can map out the way the likelihood changes with these parameters by 'brute force', meaning we compute the likelihood for many combinations of slope and intercept and plot the result. The following code generates a 50×50 array of slopes and intercepts:

```
n.b <- 50
n.a <- 50
b <- seq(0.003, 0.004, length.out=n.b)
a <- seq(-0.01, 0.025, length.out=n.a)
```

Then for each slope–intercept pair we compute the log likelihood and store this in an array

```
ll <- array(0, dim=c(n.a, n.b))
for (i in 1:n.a) {
  for (j in 1:n.b) {
    parm.ij <- c(a[i], b[j])
    ll[i,j] <- LogLikelihood(parm.ij, x, y, dy,
                             model=model.linear)
  }
}
```

R.Box 6.10
Plotting the likelihood surface

We can convert the log likelihoods to plain (linear) likelihoods and then make a fancy plot (see Figure 6.4) using e.g.

```
like <- exp(-ll)
persp(a, b, like, theta = 20, col="steelblue1",
      shade=0.5, ticktype="detailed", expand=0.5)
```

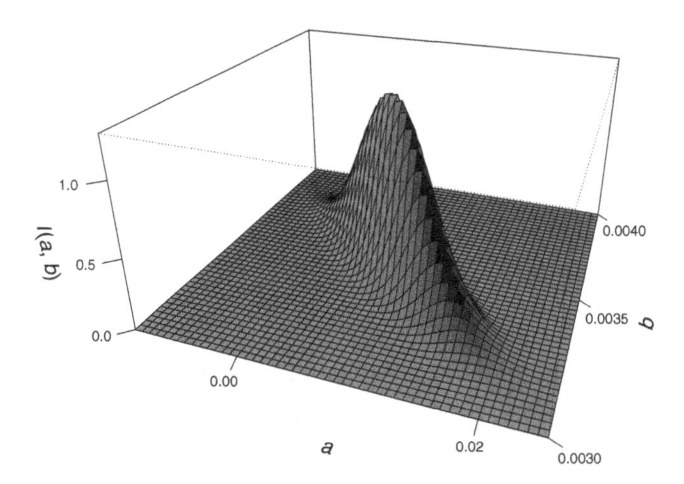

Figure 6.4 The likelihood surface for the linear model applied to Reynolds' data. The likelihood was computed at each point on a grid of parameter (a, b) values. The maximum likelihood parameters values (MLEs) correspond to the location of the peak. (The likelihood has been rescaled by a factor of 10^{-13} to reduce clutter of the axis labels.)

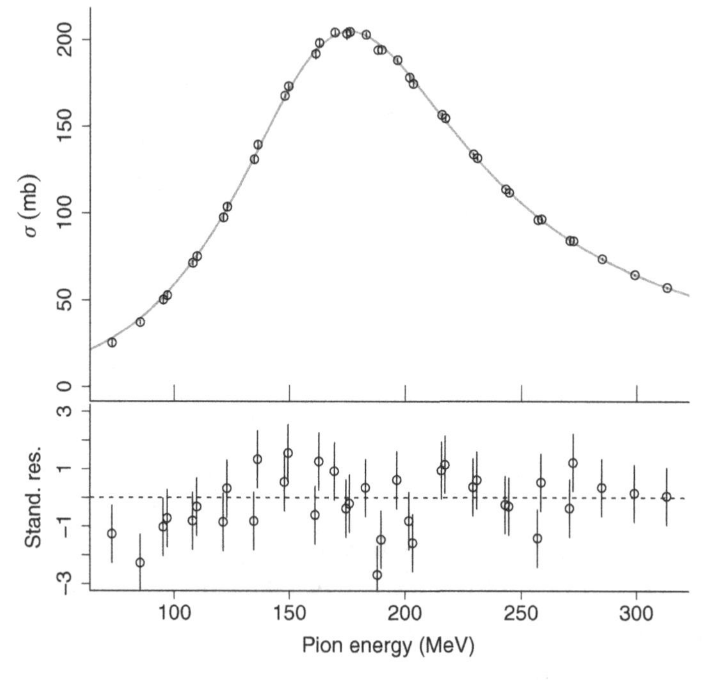

Figure 6.5 Maximum likelihood fit to the pion scattering data. The top panel shows the data (hollow circles) with errors (vertical bars) and the model (smooth curve). The bottom panel shows the standardised residuals.

6.5 Case study: pion scattering data

Section B.5 describes data from a particle physics experiment. The data comprise estimates of the scattering cross section for pion–proton interactions (the response variable) as a function of the pion kinetic energy (the explanatory variable). The response data are known to be normally distributed, and associated with each cross-section estimate is an error bar (the standard deviation). We shall use as a physical model the Breit–Wigner model formula (equation B.5). This model has three parameters: N, E_0 and Γ_0. Despite the fact that the model is more complicated, and the dataset larger than in the Reynolds example, we can find the MLEs for these parameters using exactly the same maximum likelihood method. Figure 6.5 shows the result. The MLEs for the parameters are $\Gamma_0 = 110.2$ MeV, $E_0 = 175.8$ MeV and $N = 205.0$ mb. If we use equation B.7 to convert from E to E_{cms}, we find the energy of the resonance is $E_{cms} = 1221$ MeV, within 1% of the 'textbook' value[4] (1232 MeV).

R.Box 6.11

Fitting the pion scattering data

We can fit a model to the pion scattering data by simply modifying the R code used to fit the Reynolds' data. We load the data, plot them and define the physical model as described in R.boxes B.10, B.11 and B.12. We could use the log likelihood function (R.box 6.4), but as the data are normally distributed, we shall use the chi square statistic (R.box 6.5) and minimise it as follows:

```
parm.0 <- c(100, 180, 200)
result.pion <- optim(fn=ChiSq, par=parm.0, hessian=TRUE,
                     x=x, y=y, dy=dy, model=model.pion)
mod.x <- seq(0, 400, by=1)
mod.y <- model.pion(result.pion$par, mod.x)
lines(mod.x, mod.y, lwd=3, col="red")
```

In this case, because the model is non-linear in its parameters, the minimisation (i.e. the optimisation of the parameters) can be challenging. It is especially important to make a good choice for the starting point of the minimisation. Experiment with different parameter values, plotting the corresponding models, and select those that give a model resembling the data as the starting values for the optimisation.

[4] Most of the difference is actually due to the simplifications we have made. If we subtract the background from the data and use the more complex model discussed by Pedroni *et al.* (1978), we get almost perfect agreement with the accepted value.

R.Box 6.12

Numerical MLE for the Rutherford–Geiger data

In section 6.2, we derived the analyical form of the MLE for the rate parameter, $\hat{\lambda}$, of the Poisson model used for the Rutherford & Geiger data. However, we can also solve the problem numerically as we did for the other datasets (above) by defining the minus log likelihood function and then iteratively searching for the parameter that minimises it. First we must generate a table containing the 2608 original data points (not the frequency distribution).

```
# reproduce raw data points (order does not matter)
dat <- rep(rate, freq)
plot(table(dat), bty="n")
```

Then we define the minus log likelihood function, given the data, as a function of the model's one parameter (`parm`), and search for the minimum

```
# Minus Log Likelihood fuction
LogLikelihood <- function(parm, x) {
  mlogl <- -sum(dpois(x, lambda=parm, log=TRUE))
  return(mlogl)
}
# optimise
result.ruth <- optim(LogLikelihood, par = 3.0, x = dat)
print(result.ruth$par)
```

This should give practically the same solution as using the analytical method. There will be a small difference ($\lesssim 10^{-5}$ relative difference) due to the details of the numerical optimisation; many of these details can be changed if higher numerical accuracy is needed (see `?optim`). NB: the `optim()` function may return a warning here but in this particular case it can be safely ignored.

6.6 Chapter summary

- Likelihood function (of parameters $\boldsymbol{\theta}$ for fixed data $\mathbf{y}$), assuming the data are distributed independently:

$$l(\boldsymbol{\theta}) = p(\mathbf{y}|\boldsymbol{\theta}) = \prod_{i=1}^{n} p(y_i|\boldsymbol{\theta}).$$

- Log likelihood function:

$$L(\boldsymbol{\theta}) = \log l(\boldsymbol{\theta}) = \sum_{i=1}^{n} \log p(y_i|\boldsymbol{\theta}).$$

- Weighted least squares (chi-square):

$$X^2(\boldsymbol{\theta}) = \sum_{i=1}^{n} \frac{(y_i - \mu_i)^2}{\sigma_i^2}$$

where $\mu_i = \mathrm{E}[y_i] = f(x_i, \boldsymbol{\theta})$ is the model for the response variable y.
- Ordinary least squares (OLS) when $\sigma_i = \sigma$:

$$X^2(\boldsymbol{\theta}) = \frac{1}{\sigma^2} \sum_{i=1}^{n} (y_i - \mu_i)^2.$$

- Connection between least squares and maximum likelihood:

$$X^2(\boldsymbol{\theta}) = -2L(\boldsymbol{\theta}) + const \Longleftrightarrow l \propto \exp(-X^2/2)$$

is valid only where the data are normally distributed.
- The steps in maximum likelihood model fitting are the following.
1. Obtain data y_i (perhaps taken at values x_i).
2. Construct a physical model that predicts the expected data $\mathrm{E}[y_i] = f(x_i, \boldsymbol{\theta})$, containing parameters $\boldsymbol{\theta} = \{\theta_1, \ldots, \theta_M\}$. The parameters whose values are not known *a priori* are to be estimated by fitting. The model should be based on the physics of the system we are observing.
3. Construct a probability model that describes how the data should be distributed about the physical model, $p(y_i|\boldsymbol{\theta})$. This should depend on the details of the data collection process.
4. Find the mode of the log likelihood function $L(\boldsymbol{\theta}) = \sum_i \log p(y_i|\boldsymbol{\theta})$, i.e. the parameter values that are the solutions of $\partial L(\boldsymbol{\theta})/\partial\theta_j = 0$ for each free parameter j. (When the data are normal one may equally well minimise the function X^2.) This gives the maximum likelihood estimates (MLEs) of the parameters: $\hat{\theta}_j$.

7
Significance tests and confidence intervals

About thirty years ago there was much talk that geologists ought only to observe and not theorize, and I well remember someone saying that at this rate a man might as well go into a gravel-pit and count the pebbles and describe the colours. How odd it is that anyone should not see that all observation must be for or against some view if it is to be of any service!

Charles Darwin
(letter to Henry Fawcett, 18 September 1861)

How do we know if the model fitted to our data is actually a good match to the data? And how do we quantify the uncertainty on the estimates of the model's parameters? The first question can be addressed by *significance testing*, and the second can be answered using *confidence intervals*.

7.1 A thought experiment

We shall return to the thought experiment begun in Chapter 5, drawing from a bag containing sweets of two colours, red and green. But now let us imagine that we do not know the proportions of red and green sweets. Instead, we are allowed to draw 10 times from the bag, with replacement. A simple hypothesis is that the bag contains equal numbers of red and green. What do we say about this hypothesis if we get eight greens from our 10 draws?

Let's assume the bag contains equal proportions, and find the probability for getting data like ours. As this is a situation involving a random experiment with two outcomes, the binomial distribution (section 5.2.1) is appropriate; in this case our hypothesis is that $\theta = 0.5$. This is a probability model. Now, what we need is to compute the probability for finding such surprising data, i.e. a high number of greens. Using this we can easily find that the probability of drawing eight or more green sweets out of 10 is 0.0547 (by summing the probabilities for drawing eight, nine or 10 green sweets). This quantifies how surprising it is to observe data as extreme as ours (i.e. eight or more green sweets) assuming a simple hypothesis (equal probability of drawing red and green).

Although small, this probability is not so small that we would be able to confidently rule out the 'equal proportions' hypothesis – the event we observed is not so unlikely assuming this hypothesis. If we repeated the 10-draw experiment many times, we would expect to see eight or more greens occurring in approximately 5% of repeats. Given the data our best guess of the proportion of green sweets in the bag is 80 : 20 (this is the ratio seen in our small sample), but we cannot reject the 50 : 50 ratio with much confidence.

Now let's see what happens if we had drawn nine greens and only one red sweet. The probability of observing nine or more greens is 0.011, now we may start to feel our observation is quite unlikely on the assumption of equal proportions, so may reject the hypothesis on this basis. The probability of drawing 10 out of 10 greens assuming equal probabilities is 0.000 98, a probability that is so small that we would rather reject the assumption of equal proportions than accept this as a highly unlikely outcome.

7.2 Significance testing and test statistics

Notice the structure of the above argument. We first propose a simple, and rather uninteresting, hypothesis. Then we make a statistical comparison between our observation (of a finite sample) and the predictions of the hypothesis. The observations are reduced to a statistic (in this case the number of green sweets), which would take on extreme values if the hypothesis were false. Then we find where the observed value of the statistic falls in the distribution of values expected assuming the hypothesis. If the observed value is in the far tail of the distribution – where observations are unlikely – we reject the null hypothesis. (This is a statistical analogue to the logician's *reductio ad absurdum* form of argument.) Figures 7.1 and 7.2 illustrate these ideas.

One of the main aims of statistical data analysis is to reach a decision about how well the observed data match the prediction of a theoretical model or hypothesis. A standard tool to guide such decision making is the *significance test*. The hypothesis under consideration is usually called the *null hypothesis*, denoted H_0, and we assume this is true until there is strong evidence to the contrary. Given some data we use a probability model to assess whether they are surprising. We effectively imagine a population of possible data (as if we repeated the experiment many times) assuming the null hypothesis, and assess whether our real data stand out. If the data are very surprising – in the sense that they stand out from the predictions of the null hypothesis in a way that has a low probability – then we have two choices: we just happened to get a rather improbably extreme dataset, or the null hypothesis is wrong.

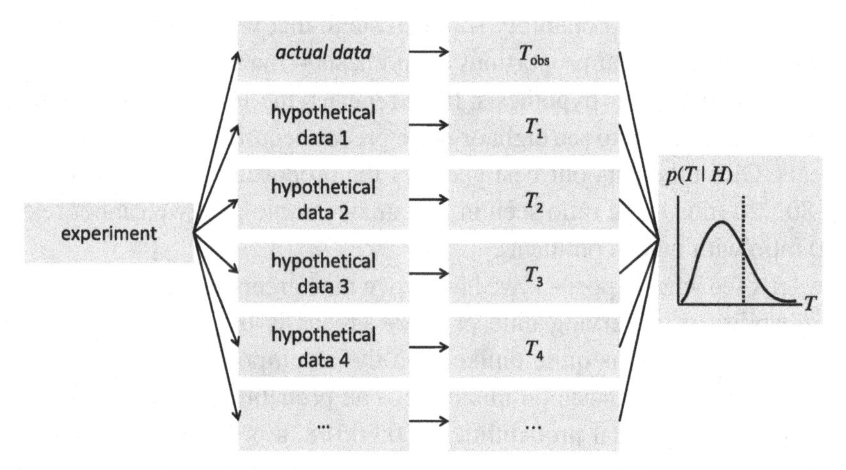

Figure 7.1 Illustration of the concepts involved in significance testing. An experiment is performed (left), some data are collected (second column). The data are processed into a single statistic T (third column). If the experiment were repeated many times, each replication would produce different random data, and therefore a different value for the statistic. The statistic therefore has some distribution (right). To perform a significance test, the observed value is located in this distribution (which can be calculated once we assume some model for the experiment).

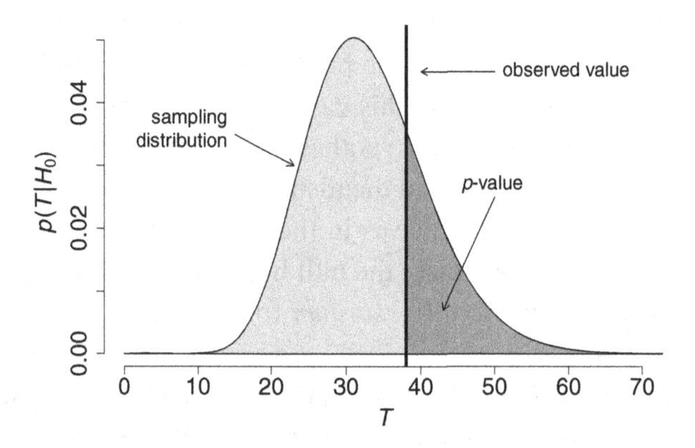

Figure 7.2 Illustration of the distribution for the test statistic T on the assumption of the null hypothesis H_0, i.e. $p(T \mid H_0)$. The observed value of the test statistic is T_{obs}. The probability of observing a test statistic as extreme or more extreme than this is p, the area under the distribution for $T > T_{obs}$ (dark grey). This is the p-value of a significance test.

We perform a significance test by first constructing a *test statistic*, which is a function of the data, $T(\mathbf{x})$. We choose the statistic $T(\mathbf{x})$ such that it is a useful measure of the overall (dis)agreement between the data and the hypothesis. Given a null hypothesis H_0 we need to calculate the probability distribution $p(T \mid H_0)$.

The outcome of a significance test is a p-value, which represents the probability of obtaining a test statistic as or more extreme than observed, if H_0 is true.

$$p = \Pr(T \geq T_{\text{obs}} | H_0) = \int_{T_{\text{obs}}}^{\infty} p(T | H_0) \mathrm{d}T. \tag{7.1}$$

Usually the resulting p-value is called the *observed significance* or the *confidence level*. The meaning of the above equation is illustrated in Figure 7.2. Here's the general idea.

1. Define the null hypothesis H_0.
2. Define a test statistic $T(\mathbf{x})$ whose sampling distribution can be calculated assuming H_0, i.e. $p(T | H_0)$.
3. Calculate the observed value of the test statistic $T_{\text{obs}} = T(\mathbf{x}_{\text{obs}})$.
4. Calculate $p = \Pr(T \geq T_{\text{obs}} | H_0)$ using $p(T | H_0)$.

This process essentially uses the data to compute a value for the random variable p, which is itself a transformation of the test statistic. If the null hypothesis is true, then p will be distributed uniformly in the interval [0, 1]. As such, p is used as a gauge of how well (or how badly) the data match the hypothesis. If p is not very small, then the observed value of the test statistic is unsurprising. A very small value of p indicates that the observed value of the test statistic is 'extreme', i.e. there is an unexpectedly large data–model mismatch, on the assumption that the null hypothesis is true. If p is small the observed test statistic value, T_{obs}, is in the far tail of the distribution of Figure 7.2. This is sometimes called a *goodness-of-fit* test, but perhaps badness-of-fit would be more appropriate.

Notice there is in general no requirement to state an alternative hypothesis; significance testing is not designed to select between alternative hypotheses, but to provide an indication of the strength of evidence against the hypothesis H_0. A high value for p means the test does not distinguish between the data at hand and the predictions of the null hypothesis: not that the null hypothesis is correct, but that there is no strong evidence to reject it. There are no special values for a threshold on p, but certain values tend to be used in science as guideline criteria (Figure 7.3) for rejecting the null hypothesis: $p \leq .05$, $p \leq .01$ or $p \leq .001$. (These correspond, approximately, to the 2σ, 2.5σ and 3.3σ tails of a normal distribution; see Figure 5.5.) We shall return to this issue in section 7.4.

If the null hypothesis is uniquely specified, i.e. it has no adjustable parameters, then it is said to be a *simple hypothesis*; if there are adjustable parameters, it is said to be a *composite hypothesis*. In the case of a composite hypothesis, the parameters need to be estimated using the data (the subject of the previous chapter).

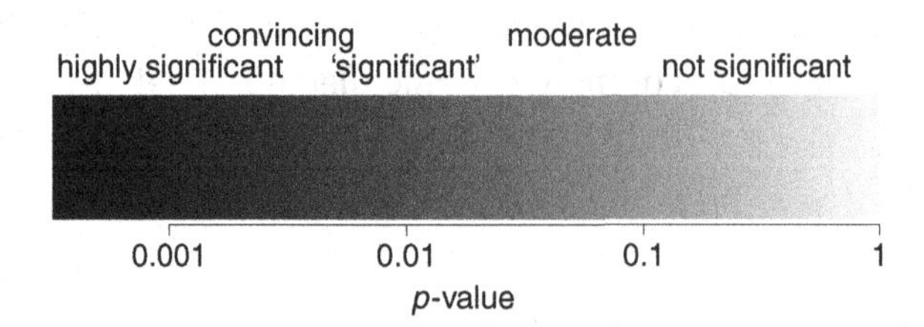

Figure 7.3 Interpreting a p-value.

7.3 Pearson's χ^2 test

Among the most popular significance tests is Pearson's χ^2 test. As discussed in section 6.4 it is often possible to employ least-squares fitting to find the maximum likelihood estimates for the parameters of a model. The most important condition that must be met for this to be valid (i.e. that $X^2 \approx -2\log l$) is that the data are distributed normally about the expected values. In this case we examine the function (equation 6.12)

$$X^2_{\text{min}} = \sum_{i=1}^{n} \frac{(\textit{observed} - \textit{expected})^2}{\textit{variance}}$$

$$= \sum_{i=1}^{n} \frac{(y_i - \mu_i)^2}{\sigma_i^2} = \sum_{i=1}^{n} \Delta_i^2. \tag{7.2}$$

The $\Delta_i = (y_i - \mu_i)/\sigma_i$ are the standardised residuals. Here y_i is the ith data point out of n data points, $\mu_i = \mathrm{E}[y_i] = f(x_i, \boldsymbol{\theta})$ is the model prediction for the ith data point calculated using parameter(s) $\boldsymbol{\theta} = \{\theta_1, \ldots, \theta_m\}$ and σ_i is the standard deviation expected for the ith data point. The parameter values that minimise this statistic, $X^2_{\text{min}} = X^2(\hat{\boldsymbol{\theta}})$, are the maximum likelihood estimates (MLEs) $\hat{\boldsymbol{\theta}}$. The model with its MLEs forms our null hypothesis H_0. Notice that X^2_{min} is a statistic; it is a function of the data only (since the parameters are obtained by maximum likelihood, or equivalently weighted least-squares fitting).

R.Box 7.1

Calculating p-values from the X^2_{min} statistic

The following examples follow directly from the previous chapter. They can be used to complete the analysis of the Reynolds' data (R.boxes 6.2–6.8) or the similar analysis of the pion scattering data (R.box 6.11).

If we have response data stored in `y`, the corresponding uncertainties stored in `dy`, a model given by the function e.g. `model.linear()` and the best-fitting (i.e. maximum likelihood) parameters stored in e.g. `result.reyn$par` or `result.pion$par`, with these in memory we can compute the X^2_{obs} statistic explicitly (equations 6.12, 7.2) or use the `ChiSq()` function (R.box 6.5):

```
X.min <- sum( (y - y.mod)^2 / dy^2 ) # ...or...
X.min <- ChiSq(result.reyn$par, x, y, dy,
               model=model.linear)
```

The number of parameters in the model, and the number of data points, can be found by querying the appropriate arrays. Given these we can compute the 'tail area' probability in the chi-square distribution with $\nu = n - m$ degrees of freedom.

```
M <- length(result.reyn$par)
N <- length(y)
df <- N - M
p <- pchisq(X.min, df=df, lower.tail=FALSE)
```

The function `pchisq()` computes the integrated probability from the chi-square probability density function. The `lower.tail=FALSE` requests that the integral above X^2_{obs} be calculated, i.e. the upper tail of the distribution (see Figure 7.2).

R.Box 7.2

Visualising the sampling distribution

If you so wish, you can visualise the pdf of X^2_{min} (i.e. the χ^2 distribution with the correct degrees of freedom ν), and the observed value X^2_{obs} relative to this, as follows:

```
x.pdf <- seq(0, df*3, by=0.1)
y.pdf <- dchisq(x.pdf, df=df)
plot(x.pdf, y.pdf, type="l", bty="n")
abline(v=X.min, lwd=4)
```

However, our analysis need not end with finding the MLEs, as X^2_{min} has another useful property. If the model is correct and the 'true' parameter values are $\boldsymbol{\theta}_0$, then the data y_i will be randomly distributed as a normal distribution with means $\mu_i(\boldsymbol{\theta}_0)$ and variances σ_i^2. The *standardised residuals* Δ_i should then have a standard normal distribution $N(0, 1)$ – i.e. zero mean and unit variance. Now, if we look again at equation 7.2, we will see that X^2 is nothing more than the sum of the squares of these standardised residuals. In section 5.3.2 we discussed the chi-square distribution and how a chi-square variable with ν degrees of freedom is formed by the summing the squares of ν independent, normally distributed random variables. This is just what we have done to calculate X^2.

This means that if our hypothesis is right – i.e. we have the right model and the correct parameter values – then X^2_{min} will have a chi-square distribution, $X^2_{\mathrm{min}} \sim \chi^2$. The function for $p(X^2_{\mathrm{min}}|H_0)$ is the chi-square function (equation 5.32). We can use X^2_{min} as a test statistic in a significance test. Using the known properties of the chi-square distribution, we can quickly calculate the p-value (equation 7.1) corresponding to any given X^2_{min}. We only need to know one more piece of information to make this possible: ν, the degrees of freedom for the appropriate distribution.

In the above argument, we assumed we knew the correct parameter values $\boldsymbol{\theta}_0$, but of course we don't; we have to make do with our estimates $\hat{\boldsymbol{\theta}}$. With this in mind, it turns out that the number of degrees of freedom is equal to the number of independent data points, n, minus the number of adjustable parameters m.

$$\nu = n - m. \tag{7.3}$$

(This may remind you of the $1/(n-1)$ in the definition of the sample variance, section 2.4.) The number of degrees of freedom is the number of independent data points minus the number of model parameters we had to estimate using the data. We now know the distribution of X^2_{min} assuming H_0.

7.3.1 Case study: pion scattering data

Let's apply this. Figure 6.5 shows $n = 36$ data points from the pion scattering experiment (section B.5). The X^2 statistic was minimised to find the best-fitting values for the $m = 3$ parameters, giving $X^2_{\mathrm{obs}} = 38.01$. This means the fit has $\nu = n - m = 33$ degrees of freedom. Comparing X^2_{obs} with the χ^2 distribution ($\nu = 33$) we can calculate the p-value to be $p = 0.25$ (see R.box 7.1). The χ^2 distribution is shown in Figure 7.2, and the area under the curve above the observed value is the p-value.

We interpret p-values as follows: in repetitions of the experiment and analysis, the probability of observing a X^2_{min} that is as extreme, or more extreme, than the observed value X^2_{obs} is 25%, assuming our model is true. In other words, the data are not surprising if we assume the null hypothesis is true. This does not mean the (null hypothesis) model is right; it means we have no reason to reject it.

7.3.2 Case study: Rutherford & Geiger's alpha decay data

We can also use Pearson's goodness-of-fit test on the Rutherford & Geiger data to see whether the Poisson model (see section 6.2) does give a good fit. We can again use equation 7.2. The data are measurements of the frequency (y_i) of different rates

Table 7.1 *Rutherford & Geiger's data (x_i is the number of scintillations per interval; y_i is the number of intervals recorded at each x_i). The model predictions are μ_i, and the standardised residuals are Δ_i.*

x_i	y_i	μ_i	Δ_i	Δ_i^2
0	57	54.306	0.366	0.134
1	203	210.256	−0.500	0.250
2	383	407.026	−1.191	1.418
3	525	525.295	−0.013	0.000
4	532	508.447	1.045	1.091
5	408	393.711	0.720	0.519
6	273	254.056	1.189	1.413
7	139	140.518	−0.128	0.016
8	45	68.006	−2.790	7.783
9	27	29.255	−0.417	0.174
10	10	11.327	−0.394	0.155
>10	6	5.798	0.084	0.007
sum	2608	2608	−2.030	12.960

of scintillations (x_i). We also have a model that predicts the expectation $\mu_i = \mathrm{E}[y_i]$, which is just the Poisson formula (equation 5.29) with $\lambda = 3.87$, multiplied by the number of observations ($n_{\mathrm{obs}} = 2608$). The y_i data are discrete counting variables, and so are themselves Poisson distributed (see section 5.2.2), meaning the variance expected on each data point is $\sigma_i^2 = \mathrm{V}[y_i] = \mu_i$. (Strictly, since the number of observations was not random the data follow a multinomial distribution, but the distinction is of little consequence here. For now we simply note this fact about the data.)

There is one more issue we must think about. Pearson's test relies on the data (y_i) having a normal distribution, but here we have Poisson distributed data. We do know, however, that a Poisson distribution can be approximated by a normal distribution, and the approximation is better for larger numbers of counts (section 5.2.2). Here we shall employ a well-used rule-of-thumb that this approximation is reasonable for $x \geq 5$; that is, where $\mu_i \geq 5$ the distribution of the data y_i should be approximately normal. For a $\lambda = 3.87$ Poisson distribution and $n_{\mathrm{obs}} = 2608$, the expected number of observations for each $x_i > 10$ is less than five, which could cause us problems. A simple solution is to 'pool' or 'bin up' the data: instead of comparing the data and model at each of $x_i = 11, 12, \ldots$, we sum the data for $x_i > 10$, sum the model similarly, and compare these to each other. Table 7.1 shows the data, including the 'pooled' data for $x_i > 10$.

The X^2 statistic here is $X^2_{obs} = 12.96$. We have here $n = 12$ data points, and we estimated $m = 1$ parameter (the mean decay rate λ) from the data. However, the number of observations ($n_{obs} = 2608$) was not random in the same way as the individual data points were. The sum of the y_i data must be $\sum y_i = n_{obs}$, and so we do not really have $n = 12$ independent y_i values: we have $n - 1$ – if we know the first $n - 1$ values of y_i we can compute the nth value using the fact that the data must sum to $n_{obs} = 2608$. (We needed to make this correction here since the number of observations was fixed. See e.g. section 11.2 of James [2006] for more on tests using histogram data.)

We have $\nu = n - m - 1 = 10$ degrees of freedom. Using the chi-square distribution we find $p = \Pr(\chi^2_{10} > X^2_{obs}) = 0.226$. Again, this p-value is not so small. It suggests that the data are not so surprising assuming that the model is true. We have little reason to be suspicious about the model.

R.Box 7.3

Computing X^2 using pooled data

We can compute the X^2 statistic using the 'pooled' data (and model). We do this by writing a new function `pool.data()` that takes as input the frequency distribution and the parameter(s) of the model. The function computes the Poisson model (R.box 6.1), finds the point where the model drops below $\mu = 5$ and then separately sums the data and model above this point.

```
pool.data <- function(parm, x, y){
  N <- length(x)
  mod.y <- model.pois(x, lambda=parm[1], scale=parm[2])
  mask <- (mod.y < 5)
  if (sum(mask) > 0) {
    m <- min(which(mask == TRUE))
    y.pooled <- sum(y[m:N])
    mod.y.pooled <- ppois(x[m-1], lambda=parm[1],
                          lower.tail=FALSE) * parm[2]
    x <- x[1:m]
    y <- c(y[1:(m-1)], y.pooled)
    mod.y <- c(mod.y[1:(m-1)], mod.y.pooled)
  }
  return(list(x=x, y=y, mod.y=mod.y))
}
```

The output is a *list* containing the pooled data (`x` and `y`) and model values (`mod.y`).

R.Box 7.4
Computing X^2 with pooled data

Using the `pool.data()` function we can write a replacement function for the X^2 statistic (R.box 6.5) so that it computes the pooled data (and model):

```
ChiSq.pool <- function(parm, x, y) {
  pooled <- pool.data(parm, x, y)
  X <- sum( (pooled$y - pooled$mod.y)^2 / pooled$mod.y)
  return(list(X=X, N.pool=length(pooled$y)))
}
```

Note that the output is now not a single number but a list containing two numbers: the X^2 value and the number of 'pooled' data points (`N.pool`). We may need to use `N.pool` later so it is a good idea to include it in the output list.

R.Box 7.5
Performing the goodness-of-fit test

Using the newly defined functions for computing the pooled X^2 (above), we can evaluate the value for the fit to the Rutherford–Geiger data, and compute a p-value as follows:

```
parm <- c(mean.rate, n.obs)
pool.results <- ChiSq.pool(parm, rate, freq)
chi.sq <- pool.results$X
df <- pool.results$N.pool - 2
p <- pchisq(chi.sq, df, lower.tail=FALSE)
cat("-- chisq:", chi.sq, " dof:", df, "p-value:",
    p, fill=TRUE)
```

7.3.3 *A rule of thumb*

There is an approximate way to gauge the goodness-of-fit. As discussed in section 5.3.2, the expectation of a chi-square distributed random variable is ν, and the variance is 2ν. Therefore, if we have ν degrees of freedom in our fit, we would expect $X^2 \approx \nu \pm \sqrt{2\nu}$. If you see that $X^2_{\text{min}} \approx \nu\,(=n-m)$, then the fit is usually 'okay'. This can be a handy rule of thumb when you are engaged in a lot of model fitting. Calculating the exact p-value will correctly take into account the degrees of freedom.

7.3.4 Applications

Now that we know the general scheme we can apply the χ^2-test in other settings. Imagine making multiple observations of a star you suspect to be variable. How can you test for variability? Define $H_0 =$ 'the star's brightness is constant', fit this model to the data (by finding the constant values that best fits all the data points) to find X^2_{obs}, and calculate the p-value (i.e. the goodness-of-fit). If H_0 does not provide a good fit to the data – i.e. p is small – you reject the model. Otherwise, we conclude the data are consistent with the starlight being constant. (Of course, we really reject the statistical model, which includes our assumptions about the errors on the measurements. It could be that there are systematic errors with the data that mean our statistical model is faulty, and so the data appear to show the star is variable even if it is not. It is important to check and verify assumptions about data where possible.)

R.Box 7.6
How many sigmas?

In the physical sciences, it is quite normal to hear people speak about 'three-sigma' results or 'five-sigma' results. This is usually just a different scale for representing p-values (many physicists deal mainly with normally distributed data). Given a normally distributed random variable with mean μ and standard deviation σ, one would expect to see values outside of the range $[\mu - 3\sigma, \mu + 3\sigma]$ with probability of only 0.0027 (see Figure 5.5). We might say that a significance test giving $p < .0027$ was a '3σ result', simply by relating the p-value to the two-tailed area of the standard normal distribution. The conversion from p-value to 'physicist sigmas' can be done easily using R. For example:

```
p <- 0.05
n.sigmas <- qnorm(p/2, lower.tail=FALSE)
```

The factor of $1/2$ is needed because we are interested in the area below $\mu - n\sigma$ and above $\mu + n\sigma$. The conversion from 'sigmas' to a p-value goes like this:

```
p <- 2*pnorm(n.sigmas, lower.tail=FALSE)
```

What p-value is equivalent to the 5σ threshold often adopted as the 'gold standard' of high-energy particle physics experiments?

7.4 Fixed-level tests and decisions

There is a related approach to assessing data–model mismatch, called hypothesis testing, which results in a decision to accept or reject the null hypothesis. In a significance test (section 7.2), one computes a p-value for the test statistic, and uses this to gauge the level of disagreement between data and model. A hypothesis test is rather like a *fixed-level* significance test. We define a threshold value α in advance of the test, and after performing the test the result is either to accept the null hypothesis if $p \geq \alpha$ or reject it if $p < \alpha$. A popular threshold is $\alpha = 0.01$, in which case it does not matter if $p = .009$ or $p = 10^{-7}$, they both lead to a decision to reject the null hypothesis. This fixed-threshold approach is useful as a model for decision making, and for highly automated procedures such as industrial quality control. The small probability α is called the *significance level* of the test, or sometimes the *size* of the test.

We can forget about the p-values themselves and instead think in terms of the test statistic T. A hypothesis test is done by defining a *critical region* for T. If the test statistic calculated using the observed data, T_{obs}, falls in this critical region, we reject the null hypothesis; if instead the test statistic falls in the *acceptance region* we accept it. The critical region is constructed such that the probability of T being observed in the region is a small number α assuming the null hypothesis is true (the default position).

For a *one-sided* test, where we are only interested in whether T is too high or not, the critical region is all values of T above some threshold T_{crit} such that

$$\alpha = \int_{T_{\text{crit}}}^{\infty} p(T|H_0)\mathrm{d}T. \tag{7.4}$$

In this case $T_{\text{crit}} = T_{1-\alpha}$, the $1-\alpha$ quantile of the distribution of T (see section 2.5). The hypothesis testing procedure is as follows.

1. Define a null (H_0) and an alternative hypothesis (H_1).
2. Define a test statistic $T(\mathbf{x})$ whose sampling distribution can be calculated assuming each hypothesis – $p(T|H_0)$ and $p(T|H_1)$ – and is different under each hypothesis.
3. Choose a significance level α.
4. Calculate the critical value of the test statistic T_{crit}.
5. Calculate the observed value of the test statistic T_{obs}.
6. Reject the null hypothesis if $T_{\text{obs}} \geq T_{\text{crit}}$ (and accept the alternative H_1), otherwise accept it.

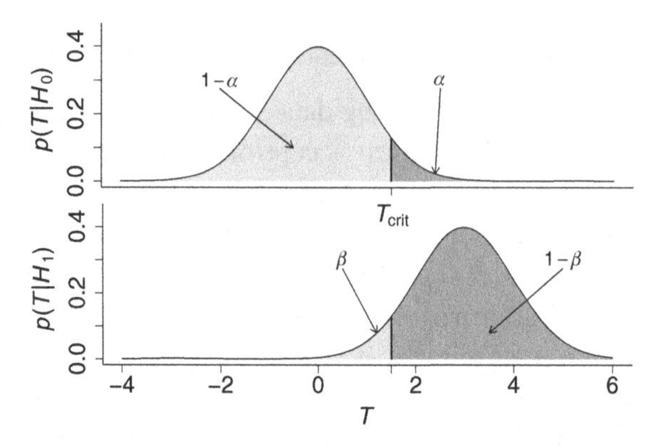

Figure 7.4 Illustration of the distributions for the test statistic T under the null hypothesis H_0 (upper panel) and alternative hypothesis H_1 (lower panel). The null hypothesis is rejected if $T > T_{\rm crit}$, which occurs with probability α assuming the truth of H_0, and probability $1 - \beta$ assuming the truth of H_1. A type I error occurs when H_0 is falsely rejected (probability α), and a type II error occurs when H_0 is falsely accepted (probability β). This is a one-sided test, since we are interested only in whether the test statistic value is too large. A two-sided test would reject the null hypothesis if the test statistic gave a value that was too large or too small.

Steps 1–4 are done before analysing the data. Then the decision to accept or reject the null hypothesis depends only on whether the observed value of the text statistic (step 5) meets the criterion given in step 6.

For a *two-sided* test, where very low or high values of T would indicate data–model mismatch, we form two critical regions. In this case the model is rejected if $T_{\rm obs} \le T_{\alpha/2}$ or if $T_{\rm obs} \ge T_{1-\alpha/2}$. Each of these should have probability $\alpha/2$ and so together they have probability α.

The other difference between significant tests and hypothesis tests is that for the latter we explicitly state an alternative hypothesis H_1 to be favoured when H_0 is rejected. This helps us choose a test statistic that is sensitive to the differences between the hypotheses, i.e. whose sampling distribution is quite different under each hypothesis (Figure 7.4). And, together with the fixed accept/reject criteria, this allows us to define the probabilities of making different types of error.

7.4.1 *Making errors*

Our decision on whether to reject the null hypothesis can go wrong in two different ways. A *type I error* is when when we reject the null hypothesis although it is in fact true (false positive); a *type II error* is when we accept the null hypothesis when it is false (false negative).

Box 7.1
Errors on trial

It might be helpful to draw an analogy with legal decisions. Consider the null hypothesis H_0: 'suspect is innocent' – this is assumed to be true (the default) unless there is good evidence to the contrary. The alternative hypothesis is H_1: 'suspect is guilty'. In these terms we might render a type I error as 'found guilty when actually innocent' and a type II error as 'found innocent when actually guilty'.

One chooses a significance level α such that the chance of making a type I error is small – numbers such as $\alpha = 0.05$, 0.01 and 0.001 are often used in practice – but there is a price to pay. Obviously every experimenter wishes to reduce the chance of making an erroneous decision. However, a smaller α means a larger $T_{\rm crit}$. This corresponds to demanding stronger evidence – a greater overall mismatch between data and model – before rejecting the null hypothesis. This increases the risk of making a *type II error*. It is of course possible that some data are well fitted by an incorrect model. An *alternative hypothesis*, H_1, will have some alternative distribution for the test statistic, $p(T|H_1)$. Given this distribution there is a finite probability that $T_{\rm obs} < T_{\rm crit}$ (leading to acceptance of H_0). This probability is given by

$$\beta = \int_{-\infty}^{T_{\rm crit}} p(T|H_1)\mathrm{d}T = 1 - \int_{T_{\rm crit}}^{\infty} p(T|H_1)\mathrm{d}T \tag{7.5}$$

and represents the probability of making a type II error. (One should choose a test statistic T for which the distributions $p(T|H_0)$ and $p(T|H_1)$ are very different, in order to maximise one's ability to distinguish the two hypotheses.) The number $1 - \beta$ is sometimes called the *power* of the test, since it gives the probability of correctly selecting the alternative hypothesis. The critical value $T_{\rm crit}$ affects both α and β. Raising $T_{\rm crit}$ decreases α but raises β. Figure 7.4 illustrates the distribution of T and the acceptance/rejection regions under both H_0 and H_1. A higher threshold makes the decision process less sensitive to correctly rejecting the null hypothesis.

Significance testing and hypothesis testing are different procedures, developed by different people[1] based on the different qualities these people thought desirable in a statistical test procedure. But both can be subject to problems of interpretation.

[1] Significance testing is almost entirely due to R. A. Fisher, although K. Pearson had proposed the χ^2 test for this purpose. E. Pearson (son of K.) and J. Neyman developed the theory of hypothesis testing.

7.5 Interpreting test results

- *Statistical significance does not mean practical significance.*
  Given a sufficient number of data almost any null hypothesis will give a poor fit, because even tiny discrepancies will be noticeable. But these may be of no practical consequence. It is up to the analyst to decide whether an effect that has been detected is physically significant.
- *Systematic errors.*
  Given very high-quality data, or a poorly calibrated experiment, *systematic errors* will become an issue. Systematic errors are biases, differences between the expected value of the data and the prediction of the physical model, that result from the experimental set-up not being perfect. If the experiment is well calibrated, and sources of contamination etc. have been minimised, the systematic errors should be small. But all experiments are prone to systematic error at some level. And if the data are very high quality, so that the statistical (random) errors are small, the systematic errors (non-random bias) may become the dominant source of data–model mismatch.
- *Lack of significance does not mean the null hypothesis is true.*
  Just because you can find a good fit, e.g. with a p-value that is not too small (e.g. $p \geq .05$), you cannot conclude that the null hypothesis is true. It is generally true that 'absence of evidence is not evidence of absence'. An acceptable fit (e.g. reasonably sized p-value) simply means the data are consistent with the model (according to your test), but there may be several different models that are also consistent with the data, and at most only one can be true. This is especially true with small datasets (or large error bars): the data may be so vague they are consistent with a wide range of models. It is also true that different statistics are more or less good at detecting different types of data–model discrepancy, so it is also possible that a good fit (accept H_0) is a result of choosing an insufficiently sensitive statistic to work with. One should chose a test statistic carefully so that it is sensitive to the kinds of data–model discrepancy indicative of an interesting alternative hypothesis (i.e. small β).

  Most important, a p-value of $p = .05$ *does not* mean that the alternative hypothesis H_1 is true with probability .95, or that the null hypothesis H_0 is true with probability .05. It says nothing about the probability of the hypotheses, just the probability of observing extreme data if the null hypothesis is true.
- *Beware of searches for significance.*
  Positive results (e.g. rejection of H_0) are systematically more likely to be written up and published compared with null results. This is known as 'publication bias' or the 'file drawer effect' because null results are filed away in researchers'

drawers. On an individual basis this might seem reasonable – you may not wish to spend time writing up an experiment that showed no interesting results – but on a more global scale this causes a problem. If we find four positive results published (each with $p \leq 0.05$) for a particular medical treatment, we might take this as good evidence in support of its efficacy. But if we also know these are the only results to have been published from a hundred trials conducted for similar treatments we would have reason to be cautious. Without further evidence it is plausible that the four positive results are all false positives (we expect $\sim 0.05 \times 100 = 5$ false positives). If we don't know how many trials were performed it can be difficult to assess the reliability of the published results.[2]

There is also a bias that can occur within a given study. For example, we might be interested in finding exceptional weather patterns using the previous year's records. We could use the data to test for extremes of temperature, humidity, wind speed, precipitation, air pressure, etc. We could also divide the data into seasons and search within each season for strong differences compared with the historical record. We might even be able to divide the data into geographical regions and examine these. We are in effect performing tests for many hypotheses (extreme temperature, wind speed, etc.) on many subsets of the data (divided by time, geography). Performing multiple hypothesis tests like this, sometimes called a 'data trawl', increases the chance of type I errors in the study (since there is a probability α *per test* of a false positive).

- *The experimenters' solution.*

Independent confirmation of results is of great importance. First results are often of borderline significance, but tentative results encourage people to perform better and more powerful experiments that provide more conclusive results. Independent experiments/observations have the added benefit of helping to reduce (or illustrate) systematic errors in any particular experiment. The reproducibility of results is very highly valued in science.

- *Beware of overfitting data.*

A 'good' fit (i.e. a fit that is not rejected in a goodness-of-fit test) is obtained once the data–model residuals are, on average, comparable in magnitude to the random errors in the data. We should not expect the model to pass through every single data point (or even every single error bar, if the data have error bars). If we use a more complex model, with more free parameters, we should be able to get an even better fit. The additional flexibility in the model means it can 'catch' every random fluctuation in the data. In such cases, where the match is too good,

[2] This can sometimes be done by analysing all the existing trial results, positive and negative, often called a *meta-analysis*.

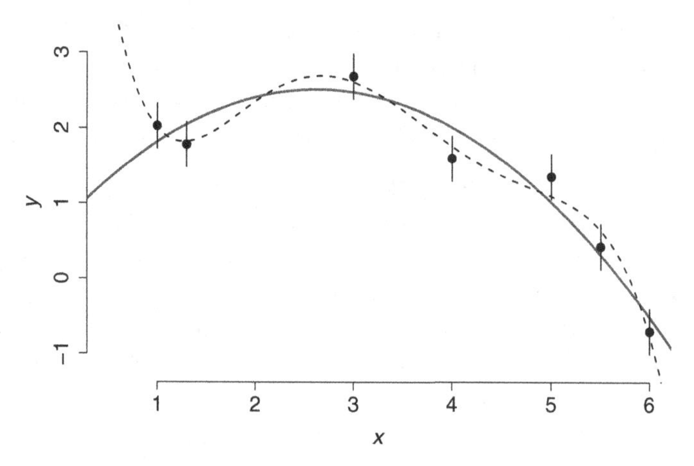

Figure 7.5 Example of overfitting. Some random data (points) were fitted using two different models: a second-order polynomial (three free parameters; solid curve) and a fifth-order polynomial (six free parameters; dashed curve).

we say the model is *overfitting* the data. See Figure 7.5 for an example. An X^2 value that is very small (or, more generally, a p-value for a goodness-of-fit test that is very close to 1, e.g. >0.99) is one sign of overfitting, i.e. that the model being used is more complex than can be justified given the data. (But this could also mean that the error bars assigned to the data points are overestimated.) Overly complex models often diverge dramatically from simpler models as we move away from the fitted data (e.g. Figure 7.5), so obtaining a few more data points outside the current range of the data can help distinguish between them.

- *Do not test a hypothesis using the same data that first suggested the hypothesis.* It is important not to mix up the exploratory and confirmatory phases of data analysis. If you have some specific hypotheses in mind, and conduct an experiment to test those hypotheses, then confirmatory analysis is what you need. If you have no particular hypothesis in mind but obtain some data in order to provide some clues about what kinds of hypothesis might work, then an exploratory analysis is what you need. But it is not legitimate to apply a statistical test to a hypothesis that was first suggested by the same data. Imagine firing bullets from a machine gun randomly into a wall, finding a cluster of five or six closely spaced bullet holes and drawing a bull's-eye around them. Now, what are the chances of all those bullets landing exactly within that bull's-eye? Hopefully you can see that this is an unfair test because the hypothesis (the bull's-eye) was designed around the data. So it is with statistical tests. In this case it is usually necessary to conduct another experiment to confirm the result suggested by the exploratory experiment.

7.6 Confidence intervals on MLEs

The process of arriving at a single number to estimate a parameter is often called *point estimation* because it summarises our inference about the parameter as a single point (a number on the real line). But this says nothing about its precision. As mentioned previously, the MLE of a parameter, being a function of random variables (i.e. experimental data), is itself a random variable and so has a probability distribution. We could do with some idea of the spread in this distribution around the expected value. The variance of the function $\hat{\boldsymbol{\theta}}$ gives us such a measure, and can be calculated using the formula for the variance of a function of random variables (equation 5.7). Notice that we are not talking about the parameter θ, which has one unique value (although unknown to us) and so zero variance. We are instead talking about our *estimate* of θ, i.e. $\hat{\theta}$. If we were to repeat the experiment several times and calculate one $\hat{\theta}$ for each experiment, they would be randomly scattered with some variance.

However, in practice it is often too difficult to solve this equation analytically. Appendix D discusses the exact theory of confidence intervals. But in practice it is common to use approximate methods based on the curvature of the log likelihood function, evaluated at the maximum likelihood estimates. It is these we discuss in the following.

7.6.1 The one-parameter case

If we have one parameter θ and random data $\mathbf{x}$ with log likelihood function $L(\theta) = \log l(\theta) = \log p(\mathbf{x}|\theta)$, then the variance of the estimator $\hat{\theta}$ is given by

$$\mathrm{V}[\hat{\theta}] \approx -\left(\frac{\partial^2 L}{\partial \theta^2}\right)^{-1}\Bigg|_{\theta=\hat{\theta}}. \tag{7.6}$$

This gives us the approximate variance of the MLEs $\hat{\theta}$ in the large-sample limit, under fairly general conditions. We will not derive this relation here, but see Cowan (1997) or James (2006) for more details, and Casella and Berger (2001) for some related mathematical proofs.[3] One way to think about this is in terms of how 'peaky' the likelihood function is. A very flat maximum of the function leads to a low value for its second derivative – a low curvature – and hence a large variance $\mathrm{V}[\hat{\theta}]$. By contrast, a very sharply defined likelihood has a large second derivative at the mode and hence a small variance.

[3] This is the Cramér–Rao lower lower bound on the variance of a parameter estimator, as applied to an *efficient* and *unbiased* MLE.

Let's take a look what this means for the Poisson example discussed above (section 6.2). We already have the first derivative (equation 6.5). Differentiating again we find that

$$V[\hat{\lambda}] \approx -\left(-\frac{\sum_i x_i}{\lambda^2}\right)^{-1}\Bigg|_{\lambda=\hat{\lambda}} = \frac{\lambda^2}{\sum_i x_i}\Bigg|_{\lambda=\hat{\lambda}} = \frac{\hat{\lambda}^2}{\sum_i x_i} = \frac{\hat{\lambda}^2}{n\hat{\lambda}} = \frac{\hat{\lambda}}{n}. \tag{7.7}$$

Recall that n is the number of observations x_i. Our estimate of the standard deviation of $\hat{\lambda}$ is $\sigma_{\hat{\lambda}} = \sqrt{\hat{\lambda}/n}$, and as we would hope the size of the interval goes down as we observe more counts. This is just like the standard error we discussed back in section 2.6. It is typical to see this presented as $\hat{\lambda} \pm \sigma_{\hat{\lambda}}$. If we use only the first $n = 20$ points from Rutherford and Geiger's data, we get $\hat{\lambda} = 3.45 \pm 0.42$ (see Figure 6.1), but if we use the full $n = 2608$ we get $\hat{\lambda} = 3.87 \pm 0.04$.

7.6.2 Graphical method

If we consider the log likelihood function $L(\theta)$ and expand this as a Taylor series around the MLE $\hat{\theta}$, we get

$$L(\theta) = L(\hat{\theta}) + \left[\frac{\partial L}{\partial \theta}\right]_{\theta=\hat{\theta}} (\theta - \hat{\theta}) + \frac{1}{2}\left[\frac{\partial^2 L}{\partial \theta^2}\right]_{\theta=\hat{\theta}} (\theta - \hat{\theta})^2 + \cdots. \tag{7.8}$$

We know that the first term in the series expansion is by definition $L(\hat{\theta}) = L_{\max}$. The second term is zero because the first derivative (the 'score function') is zero at the maximum of L. Therefore, the second-order approximation to the log likelihood is

$$L(\theta) \approx L_{\max} + \frac{(\theta - \hat{\theta})^2}{2}\left[\frac{\partial^2 L}{\partial \theta^2}\right]_{\theta=\hat{\theta}}. \tag{7.9}$$

By comparison with equation 7.6 we may write

$$L(\hat{\theta} \pm \sigma_{\hat{\theta}}) \approx L_{\max} + \frac{([\hat{\theta} \pm \sigma_{\hat{\theta}}] - \hat{\theta})^2}{2}\left[\frac{\partial^2 L}{\partial \theta^2}\right]_{\theta=\hat{\theta}} = L_{\max} - \frac{\sigma_{\hat{\theta}}^2}{2} V[\theta]^{-1} \tag{7.10}$$

and since $V[\hat{\theta}] = \sigma_{\hat{\theta}}^2$ we get

$$L(\hat{\theta} \pm \sigma_{\hat{\theta}}) \approx L_{\max} - \frac{1}{2}. \tag{7.11}$$

What this means is the following. Our estimate of the parameter is the value that maximises the log likelihood. One standard deviation either side of this estimate corresponds (approximately) to the values at which the log likelihood has decreased by 1/2 from its maximum. This method is useful when it is too difficult or too time consuming to calculate the derivatives needed for equation 7.6, or when the

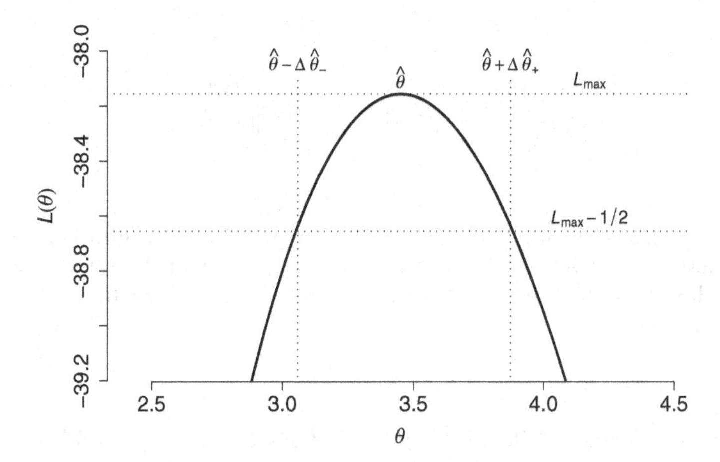

Figure 7.6 A log likelihood function $L(\theta)$. The maximum is indicated $L_{\max}$ at the MLE $\hat{\theta}$. In the large sample limit the interval $[\hat{\theta} - \Delta\hat{\theta}_-, \hat{\theta} + \Delta\hat{\theta}_+]$ corresponds to the 68.3% confidence interval ($\pm 1\sigma$).

confidence interval is not symmetric about the MLE. We find, by *brute force*[4] if necessary, the locations where the log likelihood has dropped by 1/2.

If the likelihood curve is not symmetric about its mode, we may still use equation 7.11 to construct approximate confidence intervals. Instead of corresponding to a symmetric interval $\hat{\theta} \pm \sigma_{\hat{\theta}}$ we have the interval $[\hat{\theta} - \Delta\hat{\theta}_-, \hat{\theta} + \Delta\hat{\theta}_+]$, where $\Delta\hat{\theta}_-$ and $\Delta\hat{\theta}_+$ are the distances either side of the mode where the log likelihood has dropped by 1/2.

For example, the likelihood function shown in Figure 6.1, using just the first 20 of the Rutherford & Geiger data points, gives the log likelihood shown in Figure 7.6, with $\hat{\lambda} = 3.45$. By brute force evaluation we can find the positions on the curve where the log likelihood has dropped to $L_{\max} - 1/2$, which are in this case are $[3.06, 3.87]$. These correspond to $\Delta\hat{\lambda}_- = 0.39$ and $\Delta\hat{\lambda}_+ = 0.42$. The standard deviation worked out using equation 7.6 is $\sqrt{3.45/20} = 0.42$. We can see that in this case the asymmetry in the confidence interval is rather slight, and in fact $\sigma_{\hat{\lambda}} \approx \Delta\hat{\lambda}_- \approx \Delta\hat{\lambda}_+ \approx 0.4$.

7.6.3 Coverage

We now have an estimate of our model parameter (Chapter 6) and an interval around it (using 7.11). But how do we interpret these? As the size of the data sample increases (i.e. as $n \to \infty$) the MLE $\hat{\theta}$ is distributed with a normal pdf

[4] By 'brute force' we mean performing the likelihood calculation for a set of parameter values, and using the results to map out the likelihood as a function of the parameter.

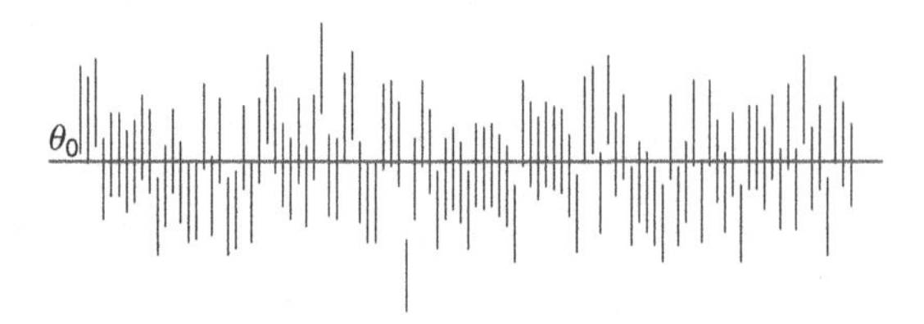

Figure 7.7 Demonstration of the coverage of the 68.3% confidence interval. 100 intervals, computed from 100 random datasets, with the same true value θ_0 are shown. Approximately 68% of the intervals 'cover' the true value (thick grey line), as expected.

centred on the true value θ (see e.g. Chapters 7 and 10 of Casella and Berger 2001). The confidence interval constructed from the two points at which the log likelihood has dropped by 1/2 from its maximum (equation 7.11) gives approximately one standard deviation either side of the mode. The probability contained within the central $\pm 1\sigma$ of a normal curve is 68.3%. This means that if we were to repeat the experiment a large number of times and calculate an interval for each, in approximately 68.3% of cases the interval would include the true value. When the interval does include the true value we say it 'covers' the true value, and in the $\pm 1\sigma$ case the *coverage* is approximately 68.3%.

Notice that we say nothing about the probability of the true parameter value θ being in the interval $[\hat{\theta} - \Delta\hat{\theta}_-, \hat{\theta} + \Delta\hat{\theta}_+]$, since the true value is not random. Instead, it is the interval that is random, since it is computed as a function of random data, and so we speak of the probability that the interval covers the true value. Since the probability contained within the central ± 1 standard deviation of a normal is 0.683, and the probability contained within the central ± 2 standard deviations is 95.3%, we say that $[\hat{\theta} - \sigma_{\hat{\theta}}, \hat{\theta} + \sigma_{\hat{\theta}}]$ is the 68.3% *confidence interval* and $[\hat{\theta} - 2\sigma_{\hat{\theta}}, \hat{\theta} + 2\sigma_{\hat{\theta}}]$ is the 95.4% confidence interval, with coverages of 68.3% and 95.4%, respectively. This approximation works in the 'large-sample' limit, when we have enough data that the likelihood function has a shape like a normal distribution.

The coverage property of the 68.3% confidence interval is illustrated in Figure 7.7.

Box 7.2

The connection between hypothesis tests and confidence intervals

A two-sided hypothesis test with a significance level of α will reject the hypothesis H_0 $(\theta = \theta_0)$ exactly when the $1 - \alpha$ confidence interval on θ does not include the value θ_0. (See also Appendix D.)

7.6.4 The many-parameter case

If our model has several parameters, $\boldsymbol{\theta} = \{\theta_1, \ldots, \theta_m\}$, we may use the generalisation of equation 7.6, which gives the *covariance matrix*. This is an $m \times m$ matrix, the elements of which are the variances of each parameter estimator and covariances between pairs of parameter estimates. To compute this we first compute the *Fisher information matrix*, which is the $m \times m$ matrix of all the second derivatives of the log likelihood function:

$$\hat{I}_{ij} = -\left.\frac{\partial^2 L}{\partial \theta_i \partial \theta_j}\right|_{\theta = \hat{\theta}}. \tag{7.12}$$

(A matrix of second derivatives is often called a *Hessian* matrix, so the Fisher information is the Hessian of the 'minus log likelihood' function.) The covariance matrix is the inverse of the Fisher information matrix

$$V_{ij} = \left(\hat{I}^{-1}\right)_{ij}. \tag{7.13}$$

This is the many-dimensional analogue of equation 7.6. It is important to note that the element V_{ij} is not the reciprocal of the element $\hat{I}_{ij}$; it is instead the element ij of the matrix inverse $\hat{I}^{-1}$. Once we have computed $\hat{I}$, and found its inverse, we have the covariance matrix (section 5.1.3). If the covariance between two parameter estimates is zero then the estimates are independent of each other. For one parameter there is only one element, the same as equation 7.6.

Now, this may seem like an awful lot of effort just to get confidence intervals on a few parameters, especially if you are not a fan of matrix algebra! But this method turns out to be more immediate than it might at first appear. Many of the standard computer algorithms for minimisation (which we might use to minimise the minus log likelihood) calculate the first and second derivatives of the function as part of the minimisation process. It is fairly standard for such routines to return not only the minimum values of the function, and parameters that give the minimum, but also the matrix of second derivatives (Hessian), from which we can calculate all the variances. This is just a short step away from the covariance matrix we want.

Once we have computed all the elements of the square matrix V_{ij} it is simple to pull out the variances of each parameter from the leading diagonal (top-left to bottom-right). Notice that the matrix is symmetric since $V_{ij} = V_{ji}$ (recall from the definition of covariance, equation 5.12, that $cov(x, y) = cov(y, x)$).

Box 7.3

Fisher information

Fisher information is a way of measuring the amount of information contained by random variables (data) about the unknown parameters of a model. If we have one parameter θ and random data $\mathbf{x}$ with log likelihood function

$L(\theta) = \log l(\theta) = \log p(\mathbf{x}|\theta)$ then the Fisher information is

$$I(\theta) = E\left[\left(\frac{\mathrm{d}L(\theta)}{\mathrm{d}\theta}\right)^2\right]. \tag{7.14}$$

(This is, by definition, the variance of the score function.) Given certain conditions (known in the trade as *regularity conditions*)

$$I(\theta) = -E\left[\frac{\mathrm{d}^2L(\theta)}{\mathrm{d}\theta^2}\right]. \tag{7.15}$$

When there are several parameters $\boldsymbol{\theta} = \{\theta_1, \ldots, \theta_m\}$ the Fisher information is an $m \times m$ matrix with elements

$$I(\boldsymbol{\theta})_{ij} = -E\left[\frac{\partial^2 L(\boldsymbol{\theta})}{\partial\theta_i \partial\theta_j}\right]. \tag{7.16}$$

In many situations it is not practical to compute the expectation as given above. Instead, one usually uses the second derivative evaluated at the maximum likelihood solution i.e.,

$$\hat{I}(\boldsymbol{\theta})_{ij} = -\left.\frac{\partial^2 L(\boldsymbol{\theta})}{\partial\theta_i \partial\theta_j}\right|_{\boldsymbol{\theta} = \hat{\boldsymbol{\theta}}}. \tag{7.17}$$

This is sometimes called the *observed Fisher information.*

R.Box 7.7

Confidence intervals from the Hessian

The minimisation function `optim()`, used in R.boxes 6.6 and 6.11, has an optional argument `hessian`, which we had previously set to `TRUE`. This requests the Hessian matrix be estimated at the minimum of the minus log likelihood – in this case the second derivatives are estimated using numerical methods. The diagonal elements of the *inverse* of the Hessian are, approximately, the variances of each parameter. Fortunately, R has a powerful battery of matrix routines to do all this complicated business for us. For example

```
covar <- solve(result.pion$hessian)
errs <- sqrt(diag(covar))
```

The first line finds the inverse of the Hessian matrix (using the powerful `solve()` function). The second line takes the square root of the elements along the leading diagonal. If we were using the `ChiSq()` function rather than `LogLikelihood()` then we would use instead

```
covar <- 2 * solve(result.pion$hessian)
```

Take a look these products:

```
print(covar); print(errs)
```

R.Box 7.8
Displaying the results

We now have MLEs (the contents of `result.pion$par`) with approximate 68.3% confidence intervals (the contents of `errs`). We could summarise these using

```
for (i in 1:M) {
   cat("Parameter", i, "=", signif(result.pion$par[i], 4),
       "+/-", signif(errs[i], 4), fill=TRUE)
}
```

This loops over every model parameter and for each it displays the best fitting value, the approximate error and the ends of the corresponding 68.3% ('1 sigma') confidence interval. The function `signif()` is used to specify the number of significant figures displayed.

7.6.5 Confidence intervals in least-squares fitting

The connection between the log likelihood function and the X^2 function allows us to construct confidence intervals in an analogous manner. In the log likelihood case we found the interval spanning ± 1 standard deviation from the region either side of $L_{\max}$ at which the log likelihood had fallen to $L_{\max} - 1/2$ (see equation 7.11). From equation 6.12, i.e. $X^2(\theta) = -2L(\theta) + const$ (in one dimension), it is clear that this is equivalent to

$$X^2(\hat{\theta} \pm \sigma_{\hat{\theta}}) = X^2_{\min} + 1. \tag{7.18}$$

Therefore, once we have found $\hat{\theta}$ as $X^2(\hat{\theta}) = X^2_{\min}$ we may then find the 68.3% confidence interval about $\hat{\theta}$ using the values of θ for which $X^2(\theta)$ increases by 1 over the minimum. Similarly, the $\pm n$ standard deviation interval around $\hat{\theta}$ may be found from the values of θ at which $X^2(\theta)$ has risen n^2 over its minimum.

The above procedure works for each parameter in the multi-parameter case, as illustrated in Figure 7.8. Here a two-parameter model $y = a + bx$ is fitted to the Reynolds' data by least squares, i.e. minimising X^2. For each parameter the extremal values that give $X^2 = X^2_{\min} + 1$ define the respective confidence intervals.

It is also possible to use the covariance matrix method. Since $X^2(\boldsymbol{\theta}) = -2 \log l(\boldsymbol{\theta}) + const$ we can form the Fisher information matrix from

$$\hat{I}_{ij} = \frac{1}{2} \frac{\partial^2 X^2(\boldsymbol{\theta})}{\partial \theta_i \partial \theta_j} \bigg|_{\boldsymbol{\theta} = \hat{\boldsymbol{\theta}}}. \tag{7.19}$$

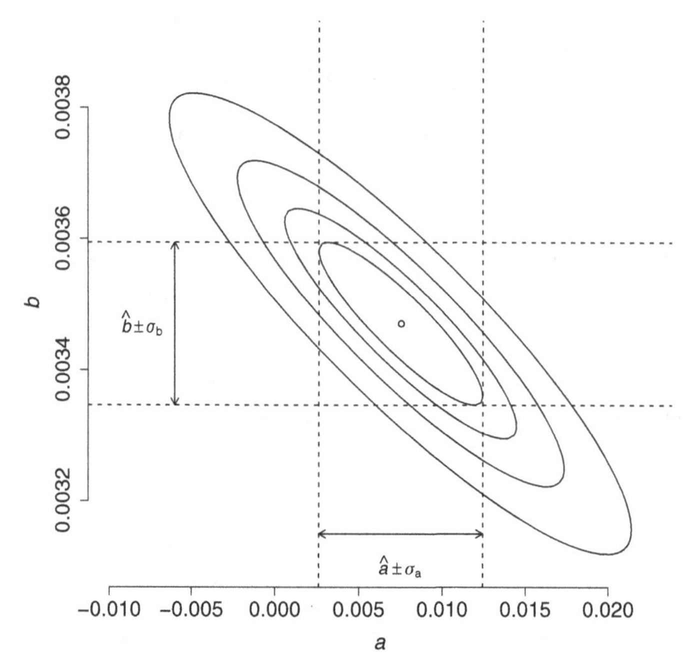

Figure 7.8 Example of 68.3% confidence intervals for two parameters of the linear model shown in Figure 3.5 (a = intercept, b = gradient). The MLEs $(\hat{a}, \hat{b})$ are indicated with the hollow circle (at the position of $min(X^2)$). The ellipses show the contours of $\Delta X^2 = 1, 2, 4$ and 8; the dotted lines indicate the corresponding 68.3% confidence regions for each of the two parameters. The fact that the ellipse is inclined (i.e. its major axis is not parallel to either parameter axis) indicates the parameters are covariant. These are effectively contours of the surface shown in Figure 6.4.

Once we have this we can compute the covariance matrix as before:

$$V_{ij} = \left(\hat{I}^{-1}\right)_{ij} \tag{7.20}$$

and extract the variances of each estimator from the diagonal elements.

7.7 Chapter summary

- Significance test for data $\mathbf{y} = \{y_1, y_2, \ldots, y_n\}$
  1. Define a test statistic $T(\mathbf{y})$.
  2. Calculate the observed value of the test statistic $T_{\text{obs}} = T(\mathbf{y}^{\text{obs}})$.
  3. Calculate the 'tail area probability' $p = \Pr(T \geq T_{\text{obs}}|H_0) = \int_{T_{\text{obs}}}^{\infty} p(T|H_0)\mathrm{d}T$.

  A small p-value is evidence against the null hypothesis.
- Pearson's χ^2 test is used to assess the goodness-of-fit of some model $\mu(\boldsymbol{\theta})$ with parameters $\boldsymbol{\theta} = \{\theta_1, \ldots, \theta_m\}$ compared with data y_i $(i = 1, 2, \ldots, n)$ using

weighted least squares. The fit statistic is

$$X^2(\boldsymbol{\theta}) = \sum_{i=1}^{n} \frac{(y_i - \mu_i(\boldsymbol{\theta}))^2}{\sigma_i^2}.$$

Once the minimum has been found, $X^2_{\min} = X^2(\hat{\boldsymbol{\theta}})$, this should follow a chi-square distribution with $\nu = n - m$ degrees of freedom (for $n > m$), assuming that the null hypothesis (i.e. the model $\mu(\boldsymbol{\theta})$, with parameters estimated by minimising $X^2(\boldsymbol{\theta})$) is true. From this it is possible to calculate a p-value. A small p-value indicates a poor match between data and model.

- In the scientific literature commonly used p-value thresholds, α, include 0.05, 0.01 and 0.001.
- Null hypothesis H_0: the hypothesis that is assumed true in the absence of evidence to the contrary.
- Alternative hypothesis H_1: the hypothesis considered as an alternative to H_0 and favoured if H_0 is rejected.
- Type I error: a false positive result, i.e. rejecting H_0 when it is in fact true.
- Type II error: a false negative result, i.e. accepting H_0 when it is in fact false.
- Hypothesis test for data $\mathbf{y} = \{y_1, y_2, \ldots, y_n\}$.
  1. Define a test statistic $T(\mathbf{y})$.
  2. Choose a significance level α.
  3. Calculate the critical value of the test statistic T_{crit} from $\alpha = \int_{T_{\text{crit}}}^{\infty} p(T|H_0)\mathrm{d}T$.
  4. Calculate the observed value of the test statistic $T_{\text{obs}} = T(\mathbf{y}^{\text{obs}})$.
  5. Reject the null hypothesis if $T_{\text{obs}} \geq T_{\text{crit}}$; otherwise accept it.

  By construction the rate of type I errors should be α.
- Coverage: the 68.3% confidence interval on a parameter $[a, b]$ is expected to 'cover' (i.e. include) the true value of the parameter θ_0 in 68.3% of repeat experiments. Likewise for the α% confidence limit. A symmetric confidence interval may be written as $[\hat{\theta} - \sigma_{\hat{\theta}}, \hat{\theta} + \sigma_{\hat{\theta}}]$ or $\hat{\theta} \pm \sigma_{\hat{\theta}}$.
- Confidence interval (68.3%) for one-parameter models from the gradient of the log likelihood

$$\sigma_{\hat{\theta}} = \sqrt{\mathrm{V}[\hat{\theta}]} \quad \text{with} \quad \mathrm{V}[\hat{\theta}] \approx -\left(\frac{\mathrm{d}^2 L}{\mathrm{d}\theta^2}\right)^{-1}\Bigg|_{\theta=\hat{\theta}}.$$

- Confidence interval (68.3%) for one-parameter models from the log likelihood function

$$L(\hat{\theta} \pm \sigma_{\hat{\theta}}) \approx L_{\max} - 1/2.$$

- Confidence interval (68.3%) for one-parameter models using least squares

$$X^2(\hat{\theta} \pm \sigma_{\hat{\theta}}) = X^2_{\min} + 1.$$

- Confidence limits on the estimator of each parameter $\hat{\theta}_i$ for many parameter models, from the covariance matrix V_{ij}:

$$\sigma_{\hat{\theta}_i} = \sqrt{V_{ii}} \quad \text{with} \quad V_{ij} = \left(\hat{I}^{-1}\right)_{ij}$$

  where $\hat{I}$ is the (observed) Fisher information matrix

$$\hat{I}_{ij} = -\left.\frac{\partial^2 L}{\partial \theta_i \partial \theta_j}\right|_{\theta = \hat{\theta}}.$$

8

Monte Carlo methods

The generation of random numbers is too important to be left to chance.
Title of an article by Coveyou (1969)

In the preceding chapters, we have discussed ways to estimate various statistics that summarise data and/or hypotheses, such as sample means and variances, parameters of models, their distributions, confidence intervals and p-values from goodness-of-fit tests. We can *calibrate* these if we know the sampling distribution of the relevant statistics. That is, we can place the observed value in the distribution expected (for a given hypothesis) and assess whether it is in the expected range or not. For example, in order to compute a p-value from a goodness-of-fit test, we need to know the distribution of the test statistics, or to find the variance (or bias) of some estimator we need to know the sampling distribution of the estimator. These follow from the distribution of the data and the mathematical relationship between the data and the statistic. Often this is difficult, sometimes even impossible, to perform analytically. But the *Monte Carlo*[1] method makes many of these problems tractable, and provides a powerful tool for analysing data, and understanding the properties of analysis procedures and experiments.

The core of the Monte Carlo method is to generate random data and use this to compute estimates of derived quantities. We can use the Monte Carlo method to evaluate integrals, explore distributions of estimators and estimate any other quantities of sampling distributions.

8.1 Generating pseudo-random numbers

When we talk about computer-generated random numbers we usually mean the output of a *pseudo-random number* generator. A truly unpredictable sequence of numbers could in principle be generated by taking recordings of radioactive decays, or atmospheric noise.[2] But in practice most Monte Carlo methods use

[1] The method was developed and named by N. Metropolis, S. Ulam and J. von Neumann, originally for simulating sequences of nuclear reactions. It is named after the famous casino.

[2] See e.g. www.random.org.

pseudo-random numbers, which are the output of a deterministic mathematical procedure chosen such that its output is unpredictable for all practical purposes. Most pseudo-random number generators (RNGs) generate a sequence of values between 0 and 1 starting from some initial state specified (at least in part) by a *seed* number. The sequence of numbers produced by starting from any particular seed should pass tests for effective randomness, but if the initial state is reset, by setting the seed number, then the same sequence will be reproduced. We shall not discuss further the details of pseudo-random number generation, as they can be found in many good books on numerical methods or scientific computing.

R.Box 8.1
Pseudo-random number generation

The simplest random number generator in R is perhaps `runif()`, which generates sequences of pseudo-random numbers uniformly distributed in the interval [0, 1]. We can use the `set.seed()` function to reset the initial state of the RNG if required. The seed only needs to be set once.

```
runif(20)         # sequence of 20 numbers
set.seed(43565) # set the 'seed'
runif(20)         # sequence of 20 numbers
runif(20)         # another sequence of 20 numbers
set.seed(43565) # reset the 'seed'
runif(20)         # same sequence as above
runif(20)         # same sequence as above
```

Notice that we did not originally specify the seed value. If no seed value is supplied it will be chosen using the computer's internal clock when required, ensuring that each time we begin a session the seed (and hence any subsequent output sequences) is be different. Once the seed is set it should not be reset unless we specifically require a repeat sequence (which can sometimes be useful for debugging purposes).

The default algorithm for pseudo-random number generation in R is the 'Mersenne Twister', one of the most popular and trusted methods. (Alternative methods are available, see `?RNG`.)

R.Box 8.2
The shape of pseudo-random numbers

We can make a crude check of the distribution of these numbers by plotting them

```
plot(runif(1000))
hist(runif(10000))
plot(runif(1000), runif(1000))
```

The distribution should be uniform, and the sequences should not show obvious structure, such as repeating patterns.

Most basic RNGs produce sequences of values uniformly distributed (section 5.3.4) in the interval [0, 1]: $u_i \sim U(0, 1)$. These can be used directly to solve some problems (e.g. Figure 8.1 and R.box 8.3) but are most often used as the basis of other pseudo-random sequences: for example, simulating random sampling from a finite set of elements. We can simulate all manner of experiments based on random samples from a finite sample space. For example, R.box 8.4 demonstrates how to solve the *birthday problem* using repeated simulation of random sampling.

R.Box 8.3

Estimating π by 'hit and miss' Monte Carlo

As a simple demonstration we estimate the value of π (the area within a circle of unit radius). This is done by randomly picking points within a square, $x \sim U(0, 1)$ and $y \sim U(0, 1)$, and computing the fraction that fall within a distance $R = \sqrt{x^2 + y^2} < 1$. As the number of points increases this fraction should converge to $\pi/4$ (see Figure 8.1).

```
calc.pi <- function(n) {
  x <- runif(n)
  y <- runif(n)
  r <- x^2 + y^2
  my.pi <- 4*mean(r < 1)
  return(my.pi)
}
```

Now that we have a function to do the relevant computation we may call it as follows:

```
result <- calc.pi(1E7)
cat(result, result-pi, fill=TRUE)
```

R.Box 8.4

The birthday 'paradox'

Given a group of N people, what is the probability that at least two of them have the same birthday? In particular, how large does N need to be for the probability of a shared birthday to be ≥ 0.5? (Assume there are 365 days in each year, and that birthdays are evenly distributed among these). This is the *birthday paradox* – it is not

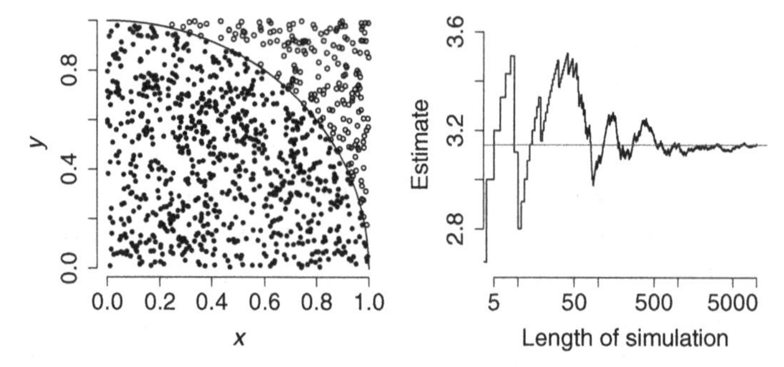

Figure 8.1 Illustration of the Monte Carlo estimate of π. Left: pseudo-random numbers x and y are generated uniformly in the interval [0, 1]. The estimate is based simply on the fraction of points that fall within the unit circle (which should be $\pi/4$ in the limit of $N \to \infty$). Right: how the estimate changes as the number of points N is increased (for one particular run of simulations).

really a paradox, but the number is smaller than most people imagine. We can solve this using the usual rules of probability. But we can also solve it using random numbers as follows

```
birthday <- function(n.sims, n.group) {
  days <- 1:365
  n.match <- array(0, dim=n.sims)
  for (i in 1:n.sims) {
    random.group <- sample(days, n.group, replace=TRUE)
    n.match[i] <- sum(duplicated(random.group))
  }
  p <- mean(n.match == 0)
  p.err <- sqrt(p*(1-p)/n.sims)
  result <- list(p=p, p.err=p.err, n.match=n.match)
  return(result)
}
```

The function `birthday(n.sims, n.group)` randomly generates `n.sims` random groups of `n.group` birthdays. (It does this using the `sample()` function to select samples of numbers between 1 and 365.) It returns the fraction of random groups for which there are no matching birthdays. An uncertainty on this is estimated using the formula for the standard deviation of a binomial variable (section 5.2.1). For example:

```
result <- birthday(1000, 23)
cat(1-result$p, "+/-", result$p.err, fill=TRUE)
hist(result$n.match, breaks=0:10-0.5, freq=FALSE)
```

8.1.1 The transformation method

Generating uniform random numbers in the range [0, 1] is useful, but we could do much more if we had a way to generate numbers with different distributions. We shall briefly consider two methods for producing random numbers with non-uniform distributions.

First, it should be obvious that given a pseudo-random sequence u_i from the $U(0, 1)$ distribution we can generate a sequence from the $U(a, b)$ distribution using $x_i = (b - a)u_i + a$. This is an example of a transformation applied to a random sequence to change its distribution. We saw another example of a transformation in section 5.4, that time from a normal to a chi square distribution. Is there a way to transform from $U(0, 1)$ values to other useful distributions?

Fortunately, there is a simple method for finding a suitable transformation in many cases. We wish to generate values of a variable X with pdf $p_X(x)$, and cdf $F_X(x)$. We can use the basic property of the cumulative distribution (e.g. equation 4.31): the distribution of the variable $u = F_X(x)$ is uniform, $U(0, 1)$. This means that if we had a variable X with density $p(x)$ we could transform it to another variable with a uniform $U(0, 1)$ distribution using $u = F(x)$. If we can find the inverse of this, then we can transform from our uniform u to our target density $p(x)$: $x = F_X^{-1}(u)$.

Box 8.1

Why the inverse transformation method works

Consider two continuous random variables: U with uniform pdf $U(0, 1)$, and X with pdf $p_X(x)$. The cdf of X we call $F_X(x)$, and that of U is simply $F_U(u) = u$ (over $0 \leq x \leq 1$). If we define $X = F_X^{-1}(U)$, we can find its cumulative distribution

$$\begin{aligned}
\Pr(X \leq x) &= \Pr(F_X^{-1}(U) \leq x) \\
&= \Pr(F_X(F_X^{-1}(U)) \leq F_X(x)) \\
&= \Pr(U \leq F_X(x)) \\
&= \Pr(0 \leq U \leq F_X(x)) \\
&= F_X(x) - 0 \\
&= F_X(x).
\end{aligned}$$

And so $X = F_X^{-1}(U)$ does have the cdf, and hence pdf, we expected.

As a simple example, let's see how to generate a variable X with an exponential distribution. Our target is

$$p_X(x) = \lambda e^{-\lambda x} \Rightarrow F_X(x) = 1 - e^{-\lambda x} \tag{8.1}$$

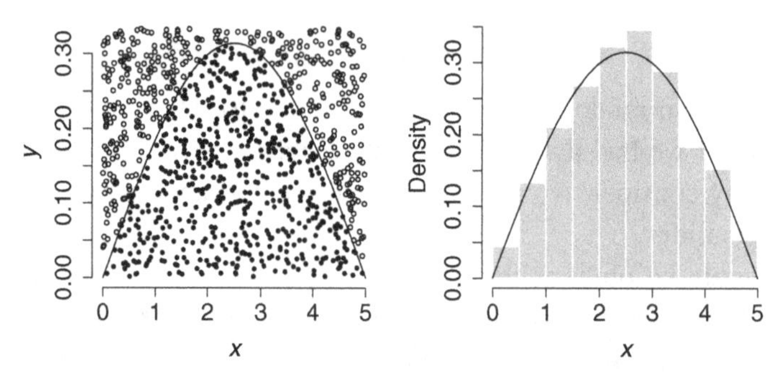

Figure 8.2 Illustration of the accept–reject method for generating pseudo-random numbers with a specified pdf. Left: the curve shows the desired pdf $p(x)$, and the points correspond to 1000 pairs of uniformly distributed values (x_i, y_i). Those falling under the curve $p(x)$, i.e. with $y_i \leq p(x_i)$, are accepted (filled symbols), whereas those falling outside are rejected (hollow symbols). Right: histogram of the x values of the accepted points.

and to do this we set $U = F_X(X)$ and solve for X

$$U = 1 - \mathrm{e}^{-\lambda X}$$
$$X = -\frac{1}{\lambda}\log(1 - U)$$

But quite clearly U and $1 - U$ both have uniform distributions, so we write the transformation from uniform to exponential pdf as

$$X = -\frac{1}{\lambda}\log(U). \tag{8.2}$$

The inverse transformation method is very powerful, but there are some distributions for which the inverse cdf cannot be found, and for these we need other methods.

8.1.2 The accept–reject method

The accept–reject method (aka hit-and-miss) is less efficient than the transformation method, in the sense that we may need to generate many more uniformly distributed random numbers than we get output numbers with the target distribution. The method is quite simple. Picture a pdf curve bounded from above by a box. If we generate random points within the box using a uniform distribution (for horizontal and vertical positions) and keep only the points that fall under the pdf curve, they will have the desired distribution. This is illustrated in Figure 8.2.

To see how this works consider the process in steps. When we draw points randomly from the box we first draw a random x_i value, then a random y_i value. We accept the x_i value if $y_i \leq p_X(x_i)$, otherwise we reject it and draw another point (x_i, y_i). The points x_i are therefore accepted with probability proportional to $p_X(x_i)$, exactly as needed to give the correct distribution.

The accept–reject method is useful in cases where the transformation method cannot be applied. The acceptance rate–the fraction of random numbers accepted – is a measure of the computational efficiency of the procedure, and depends on the shape of the target distribution and the size of the enclosing box. The method is easily generalised to allow the original points to be drawn from a non-uniform distribution $p_Y(y)$ in order to improve the efficiency (i.e. fewer rejections).

8.2 Estimating sampling distributions by Monte Carlo

Given the computational tools to generate data from the most frequently encountered distributions, we can solve all manner of data analysis problems.

8.2.1 Case study: Rutherford and Geiger data

In earlier chapters we analysed Rutherford and Geiger's data (section B.2). We used a Poisson model and found an estimate for the rate parameter λ (section 6.2) and an approximate confidence interval (section 7.6), and showed that the model provided a reasonable match to the data (section 7.3.2). We can use the Monte Carlo method to study the sampling distribution of the statistics we used. We first describe the steps of the analysis of the real data, then simulate the process in the computer.

The process of obtaining the analysis results involved several steps.

1. Collect ($n_{\text{obs}} = 2608$) observations x_i (scintillations per time interval).
2. Reduce the n_{obs} 'raw' data points into a frequency distribution, y_j (with e.g. $j = 0, 1, \ldots$).
3. Find the MLE rate ($\hat{\lambda}$), and a confidence interval, using a Poisson frequency distribution as the model.
4. Perform Pearson's goodness-of-fit test (compute X^2) for the model as fitted to the frequency distribution.

Now we can translate these into a computer routine to simulate the experiment and analysis, and then run it as many times as we like to find the sampling distribution of any statistic.

1. Repeat for each of $k = 1, 2, \ldots, n_{\text{sim}}$ simulations the following.
   - Simulate: draw n_{obs} numbers from a Poisson distribution $x_{i,\text{sim}[k]} \sim Pois(\lambda)$
   - Process: compute the frequency distribution of these data $y_{j,\text{sim}[k]}$

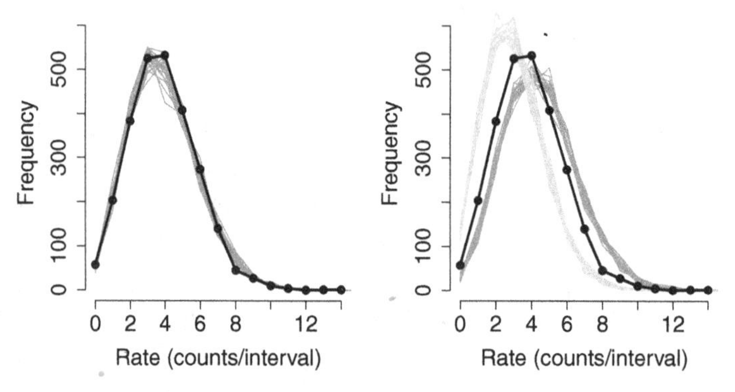

Figure 8.3 The Rutherford and Geiger data compared with some simulated datasets. The real data are shown in solid black. Left: data simulated using $\lambda = 3.87$ (the MLE for the rate parameter). Right: data simulated using $\lambda = 3.0$ and $\lambda = 4.5$. The observed data appear pretty unexceptional among the set of simulated data in the left panel, but are clearly exceptional if drawn from either set of simulated data in the right panel.

- Estimate: compute the rate estimate as for the real data $\hat{\lambda}_{\text{sim}[k]}$
- Test: compute Pearson's goodness-of-fit statistic $X^2_{\text{sim}[k]}$ (exactly as in section 7.3.2).

2. Examine the distribution of the $\hat{\lambda}_{\text{sim}[k]}$ and $X^2_{\text{sim}[k]}$ values.

The procedure generates (pseudo-) random data as if we repeated the experiment n_{sim} times, and the true rate was λ. The data point $x_{i,\text{sim}[k]}$ is the ith data point from the kth simulated experiment. For each random dataset we repeat the analyses applied to the real data, so e.g. $X^2_{\text{sim}[k]}$ is the goodness-of-fit statistic from the kth simulated experiment. We thereby obtain a sample of X^2 and $\hat{\lambda}$ values drawn from the sampling distribution, assuming the hypothesis that the distribution of the data x_i was indeed $Pois(\lambda)$. Figure 8.3 shows examples of simulated data, and Figure 8.4 shows the distributions of the statistics X^2 and $\hat{\lambda}$ based on simulations (assuming $\lambda = 3.87$). We can use these to check our analysis methods.

R.Box 8.5
Monte Carlo analysis of Rutherford and Geiger's data

We start by defining a function that will take as input the Poisson rate parameter and the number of observations to make, and produce as output the frequency distribution of the random observations.

```
sim.data <- function(lambda, N.data) {
  breaks <- 0:30 - 0.5
  data.sim <- rpois(N.data, lambda)
```

```
  hist <- hist(data.sim, breaks=breaks, plot=FALSE)
  result <- list(x=hist$mids, y=hist$counts)
  return(result)
}
```

Now let's see some results (as in Figure 8.3):

```
sim.1 <- sim.data(3.87, 2608)
plot(sim.1, type="o", bty="n", xlab="Rate",
     ylab="Frequency", main="simulation")
```

R.Box 8.6
Computing the sampling distribution

Using the function `sim.data()` we can simulate data, then analyse it as with the real data (i.e. compute the mean counts/interval and then the X^2 statistic for the corresponding Poisson model – R.box 7.4) using the 'pooled' chi-square test of R.boxes 7.3 and 7.4.

```
N.sim <- 1E4
stat.sim <- array(0, dim=N.sim)
mean.sim <- array(0, dim=N.sim)
breaks <- 0:30 - 0.5
for (k in 1:N.sim) {
  sim <- sim.data(mean.rate, n.obs)
  mean.sim[k] <- sum(sim$x * sim$y) / n.obs
  parm.k <- c(mean.sim[k], n.obs)
  fit.k <- ChiSq.pool(parm.k, sim$x, sim$y)
  stat.sim[k] <- fit.k$X
}
```

R.Box 8.7
Summaries of the Monte Carlo results

With the simulation results stored as arrays `stat.sim` and `mean.sim`, we can plot their histograms

```
par(mfrow=c(1,2))
hist(stat.sim, xlab="X.sq", col="blue", freq=FALSE)
hist(mean.sim, xlab=expression(widehat(lambda)),
     col="blue", freq=FALSE)
layout(1)
```

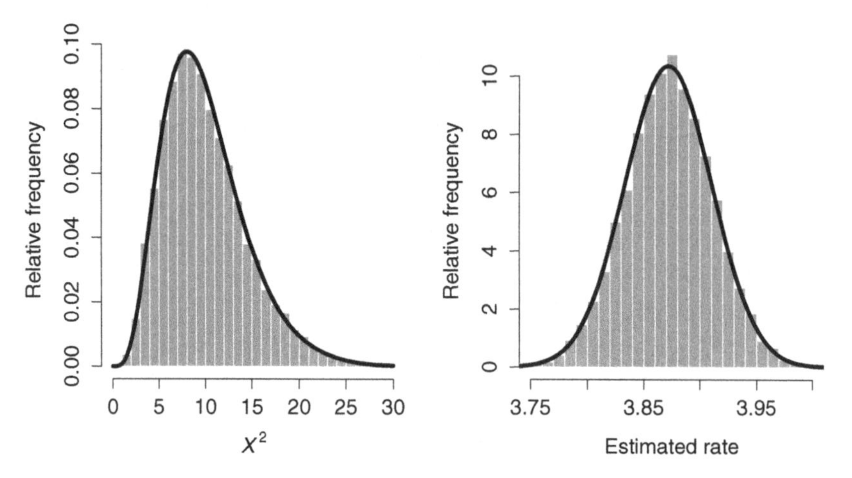

Figure 8.4 Results of $N = 10^4$ simulations of the Rutherford and Geiger dataset. Left: histogram of the statistic X^2 from each simulation. The solid curve shows the chi-square distribution with $\nu = 10$ degrees of freedom (see 7.3.2). Right: histogram of the rate estimates from each simulation (the sample means). The solid curve shows a normal distribution with $\mu = 3.87$ and $\sigma = 0.04$ (see section 7.6.1). We can see that in this case the sampling distributions of these statistics (computed assuming the null hypothesis) are close to their expected (approximate) distributions.

and find a 68.3% confidence interval ($\pm 1\sigma$) for the rate estimate:

```
print(quantile(mean.sim,  probs=pnorm(c(-1,1))))
```

and compute a p-value from the fraction of simulated X^2 values that exceed the observed value $X^2_{\text{obs}} = 12.96$ (section 7.3.2 and R.box 7.5)

```
print(mean(stat.sim > chi.sq))
```

In section 6.2 we estimated the variance on the rate estimate using an approximation: $\hat{\lambda} = 3.87$ and $\sigma_{\hat{\lambda}} = 0.04$. Based on the $n_{\text{sim}} = 10^4$ simulations shown in Figure 8.4 we can approximate the expected value and standard deviation of $\hat{\lambda}$. These are 3.87 and 0.04, respectively. These confirm that the estimator has a small bias (we got back an average $\hat{\lambda}$ very like the λ we put in), and our estimated standard deviation is accurate. Figure 8.4 (right) compares the Monte Carlo distribution of $\hat{\lambda}$ with a normal distribution with the mean and standard deviation as estimated from the real data, providing a visual confirmation that these are reasonable.

In section 7.3.2 we computed a goodness-of-fit statistic X^2, and compared this with an assumed sampling distribution (χ^2 with $\nu = 10$) to obtain a p-value. We can use the simulations to examine whether the assumed distribution does indeed match the sampling distribution of the data, and to give a Monte Carlo estimate of the p-value. Figure 8.4 (left) compares the Monte Carlo distribution of X^2 to the assumed distribution, showing that the sampling distribution of X^2 is indeed close to the assumed form. We can also compute the fraction of simulations for which $X^2_{\mathrm{sim}[k]} > X^2_{\mathrm{obs}}$, which gives 0.2283, again very close to the p-value computed assuming the χ^2 distribution.

8.2.2 Case study: pion scattering data

The general scheme outlined above easily generalises to problems involving multi-parameter models, such as fitting the pion scattering data with the Breit–Wigner model (sections 6.5 and 7.3). In that case we can use the best-fitting model as our estimate for the 'true' spectrum from which to simulate new, randomised data. This is done by adding a normally distributed random error to the model, or equivalently drawing each new data point from a normal distribution centred on the model, at the same energy positions as the original data. We can generate a large sample of simulated data and fit this to obtain sampling distributions for the parameter estimates and the chi-square statistic (e.g. Figure 8.5).

R.Box 8.8

Monte Carlo simulations of the pion data

From R.box 6.11 we should have a model fitted to the pion scattering data, i.e. the MLEs for the three parameters. We can use the Monte Carlo method to map out the joint distribution of the parameter estimates, assuming this model to be true. With the relevant `ChiSq` function defined (R.box 6.5) and the data (`x`, `y`, `dy`) and model (`mod.y`) in memory, we can generate Monte Carlo distributions for the parameter estimates and the chi square statistic as follows.

```
N.sim <- 1000
M <- length(result.pion$par)
mod.y <- model.pion(result.pion$par, x)
parm.sim <- array(0, dim=c(N.sim, M))
chisq.sim <- array(0, dim=N.sim)
for (i in 1:N.sim) {
  y.sim <- rnorm(length(mod.y), mean=mod.y, sd=dy)
  result.sim <- optim(fn=ChiSq, parm=parm.0,
                      x=x, y=y.sim, dy=dy,
                      model=model.pion)
  parm.sim[i,] <- result.sim$par
  chisq.sim[i] <- result.sim$value
}
```

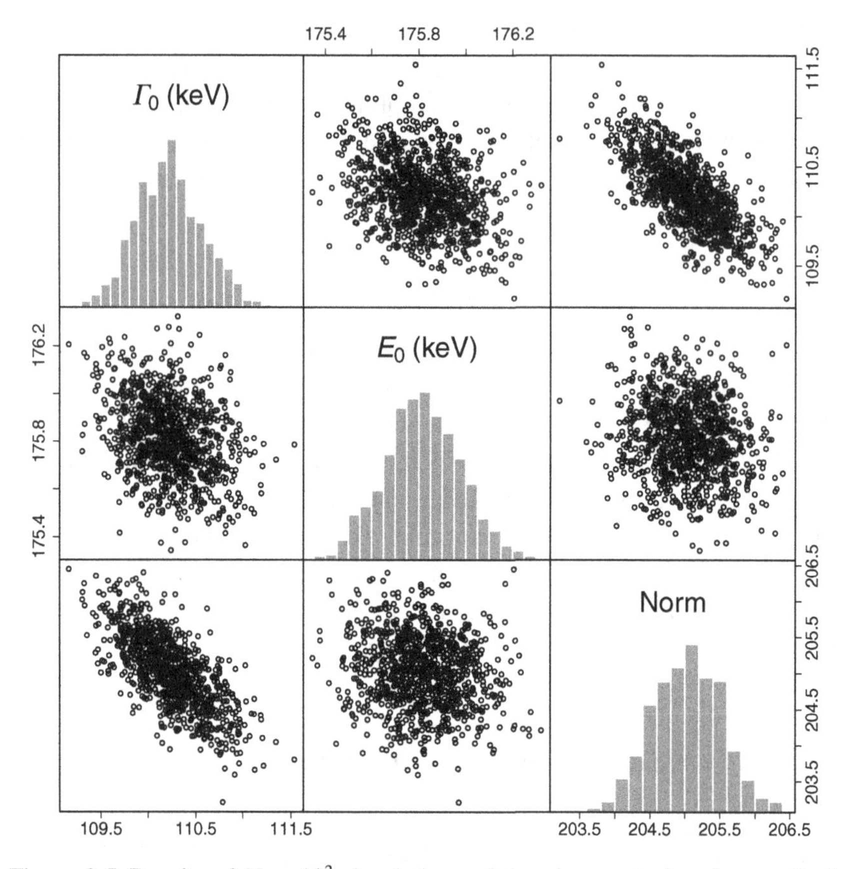

Figure 8.5 Results of $N = 10^3$ simulations of the pion scattering dataset. Each of the simulated datasets was based on the best-fitting model (section 6.5) but with randomised normal errors, and each was fitted by weighted least squares to estimate the parameters. The scatter plots show the resulting parameter estimates. The simulations make it relatively easy to examine the marginal (one parameter) or joint (multiple parameter) distributions and derive summaries from them.

This also requires that a suitable `parm.0` be defined (see R.box 6.11). The core of this is the loop that repeats the same commands `N.sim` times. Each time around the loop we generate a new set of `y` (cross-section) data values by drawing from a normal distribution whose mean is the same as the model value and whose standard deviation is the same as the error value, at each value of `energy`. These simulated data are then fitted by minimising the `ChiSq` function and the best-fitting parameters and minimum X^2 values are stored in arrays for later use. The results can be plotted using e.g. `hist(chisq.sim)` or `pairs(parm.sim)` (Figure 8.5) or summarised as in R.box 8.7.

What do we learn from this extra work? In this particular case, not a great deal! The Monte Carlo distributions of the parameter estimates are very similar to those we can infer from the covariance matrix estimated during the fitting process, and the Monte Carlo distribution of the chi-square statistic is very like that we would predict using the standard theory (i.e. a χ^2 distribution with $\nu = 33$). But that did not have to be the case. The covariance matrix gives us an estimate of the way the parameter estimates are distributed, but only in terms of the shape of a normal distribution. It could have been the case – and often is with more complicated models – that the sampling distributions are not so simple, and assuming normality could lead to spurious conclusions. Using the Monte Carlo simulations, we can 'map out' the distributions of almost any parameter estimator without having to assume a particular analytical form for its distribution.

But we can do more now we have the tools to generate and analyse Monte Carlo data. With only small modifications to these we can investigate the effects having larger or smaller errors on the data, or non-normal errors, or including more/fewer data points, etc. We could add additional filtering steps to the analysis and see how these change the quality of the final results. This information can then be used to assess the robustness of our current data and analysis, and design future experiments and analyses.

8.3 Computing confidence by *bootstrap*

The preceding sections discussed how to generate and analyse simulated data generated from different distributions. But we had to assume some particular distribution for the data generating process (e.g. Poisson for the Rutherford–Geiger data, normal for the pion scattering data). In some cases we know too little about the form of the distribution to be able to make such an assumption. So how can we use Monte Carlo methods if we do not know the distribution(s) from which our data were drawn?

There is a method called the *bootstrap* that can be used in such cases (it is an example of the class of *resampling* techniques). The essence of the bootstrap is to use the data as a model for its parent distribution. If we have a sample of data $\mathbf{x} = \{x_1, x_2, \ldots, x_n\}$, we generate a bootstrap sample $\mathbf{x}^*$ by drawing n elements, with replacement, from this list. The bootstrap sample will probably include repeated values (e.g. x_2 may appear twice, and x_1 not at all), which may seem odd, but the bootstrap sample is (probably) different from the original data, and we know the distribution of bootstrap samples is the same as the empirical distribution of the original data sample (which we take as a model of the population).

For example, we could estimate the standard deviation of the sample mean as follows. Draw a bootstrap sample $\mathbf{x}^*$, comprising n values drawn with replacement

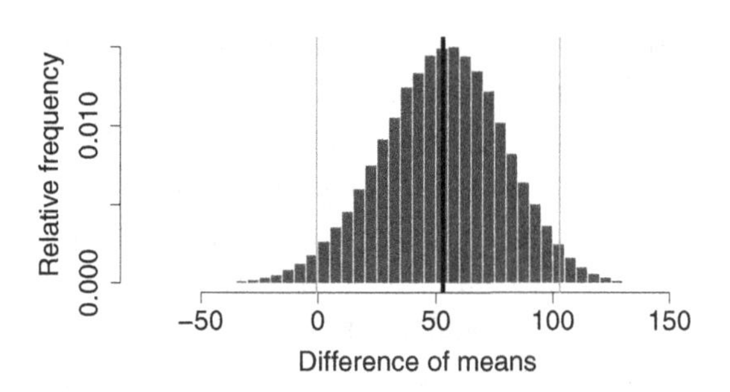

Figure 8.6 Distribution of the difference of means for 10^5 bootstrap simulations of two samples of the Michelson speed of light data. Each sample contains 20 data points (see R.boxes 3.1 and 3.3). The vertical bar near the center indicates the difference between the means for the original data, the histogram shows the distribution derived from the bootstrap samples, and the two vertical bars at the sides indicate the bootstrap 95% confidence interval (i.e. the 0.025 and 0.975 quantiles of the bootstrap distribution). This interval (just) includes the zero point.

from the original sample $\mathbf{x}$, and compute its mean, $t^* = \overline{\mathbf{x}^*}$. Then repeat for a large number of bootstrap samples. This gives us a set of values for the bootstrap sample means (t^*), from which we can compute the standard deviation.

For example, in section 3.2 we used the two-sample t-test to examine the differences between the means of two samples $\mathbf{x}$ and $\mathbf{y}$. R.box 8.9 and Figure 8.6 illustrate the bootstrap method as applied to the 'difference of two means' problem. This gives us a bootstrap estimate for the distribution of the difference of the same means. We generate a large set of bootstrap datasets $\mathbf{x}^*$ and $\mathbf{y}^*$, and compute the sample mean of each, and their difference, $t^* = \overline{\mathbf{x}^*} - \overline{\mathbf{y}^*}$. Each bootstrap dataset is a pair of samples, each one drawn randomly with replacement from the corresponding original data sample. We can use these to map out the bootstrap distribution of the statistic without making assumptions about the parent distributions of each sample.

Note that this does not assume the null hypothesis that the two samples have equal (population) means. If we wanted to form bootstrap simulations for that situation, we could do so by drawing each bootstrap sample ($\mathbf{x}^*$ and $\mathbf{y}^*$) from the joint sample $\{\mathbf{x}, \mathbf{y}\}$. From these we could estimate the p-value (for the significance test that the two samples have the same mean) as the fraction of simulations for which $t^* > t_{\text{obs}}$.

R.Box 8.9

Bootstrap study for the difference of two means

In Chapter 3 (R.boxes 3.1 and 3.3) we looked for differences between the means of two samples of data from Michelson's speed of light experiments. Following from this we can generate a bootstrap distribution for the difference in the sample means

```
N.sim <- 1E5
t.boot <- array(NA, dim=N.sim)
for (i in 1:N.sim) {
  x.boot <- sample(x, replace=TRUE)
  y.boot <- sample(y, replace=TRUE)
  t.boot[i] <- mean(x.boot) - mean(y.boot)
}
hist(t.boot)
print(quantile(t.boot, probs=c(0.025, 0.975)))
```

Figure 8.6 illustrates the output. The first two lines of code set the number of simulations to use, and define an array to store the results (the bootstrap estimates of the difference in the means). Inside the `for ...` loop we generate two new bootstrap data samples by sampling with replacement from the original data, and then computing the difference in their means.

8.4 The power of Monte Carlo

We have looked at just a few of the ways to simulate random data, and make use of those simulations. But in order to use the Monte Carlo method effectively we do need to be careful. The most obvious point is that we need to generate a sufficient number of simulated data to be confident in the results. For example, if we want to estimate the 95% confidence interval for some statistic, we must generate a lot more than 100 simulations, otherwise we will have too few simulations in the tails of the distribution to define the interval well. The *law of large numbers* means we can improve our Monte Carlo estimate by generating and using more simulations, at the expense of more computing time.

We also need to be sure that our simulations are a reasonable representation of the relevant data collection process. As with so much of computing: *garbage in, garbage out*. But often the process of developing a Monte Carlo test of some data analysis is a useful process in itself. It forces us to think what is random and what is not. For example, in Rutherford and Geiger's experiment the counts per interval were random, but the number of intervals recorded was not random. At least, we assumed it was set by practicalities of the experiment and that if the experiment were repeated the same number of records would be taken. But it's always possible to change this assumption and learn how this affects the final results.

The process of generating and analysing data is also a great test of the analysis itself. With large and costly experiments it is now quite normal to simulate a range of plausible datasets well in advance of the experiment. These can be used to test whether the experiment will produce enough data, of the right kind, to achieve its intended goals. This also tests whether the data analysis procedures are adequate. Analysing simulated data – where we know the 'true' values of any model parameters – can be a great way to debug an involved analysis procedure.

8.5 Further reading

Chapter 3 of Cowan (1997) gives a very brief review of Monte Carlo methods, which are then used later in the book. Albert (2007) gives lots of examples using R to solve statistical problems (the Bayesian way) using Monte Carlo methods. There are many books dedicated to the mathematics and computer science behind pseudo-random number generation (e.g. Gentle, 2003). Efron and Tibshirani (1993) describe the background, application and theory of bootstrap procedures.

8.6 Chapter summary

- Given pseudo-random numbers with a $U(0, 1)$ distribution, u_i, we can simulate many other distributions using
  - accept reject method
  - transformation method (including inverse-cdf transformation)
  - other specialised methods.
- We can also simulate random sampling from some discrete population.
- We can use sequences of pseudo-random numbers to integrate functions, including otherwise intractable multi-dimensional problems, with precision.
- Given pseudo-random simulations of experimental data (e.g. $\{x_i, y_i, \ldots\}$) we can 'calibrate' or 'map out' the sampling distribution of any statistical summary of the data (assuming the particular hypothesis used to generate the data). The usual applications of this are the following.
  - Calculate p-values for goodness-of-fit tests when the analytical form of the sampling distribution of the test statistic is not known or difficult to use.
  - Calculate $(1 - \alpha) * 100\%$ confidence intervals when the analytical form of the sampling distribution of the estimator is not known or difficult to use.
  - Test the 'power' of the experiment and/or analysis procedures – quantify its performance under different (theoretical) hypotheses and experimental set-ups.

Appendix A

Getting started with statistical computation

Dotted throughout the book are extracts of computer code that show how to perform the calculations under discussion. The examples are based on specific problems discussed in the text, but should be clear enough that they can also be used, with very little effort, for 'real life' data analysis problems. The computer codes are written in the R environment, which is introduced in this appendix.

A.1 What is R?

R is an environment for statistical computation and data analysis. You can think of it as a suite of software for manipulating data, producing plots and performing calculations, with a very wide range of powerful statistical tools. But it is also a programming language, so you can construct your own analyses with a little programming effort. It is one of the standard packages used by statisticians (professional and academic). To install R visit www.r-project.org/.

A.2 A first R session

First of all, start R, either by typing R at the command prompt (e.g. Linux) or double-clicking on the relevant icon (e.g. Windows).

A typical R session involves typing some commands into a 'console' window, and viewing the text and/or graphical output (which may appear in a pop-out window). The prompt is usually a '>' sign, but can be changed if desired. At the prompt you can enter commands to execute. Virtually all commands in R have a `command(arguments)` format, where the name of the command is followed by some arguments enclosed in brackets (if there are no arguments the brackets are still present but empty).

Try the following commands to demonstrate some of the graphical capabilities of R. Anything following a hash ('#') is a comment and is ignored by R.

```
demo(graphics)          # demonstrate graphics
demo(persp)             # demonstrate 3D perspective
demo(image)             # demonstrate images
example(plot)           # examples of plot() command
?plot                   # help for plot() command
```

Now let's try some simple arithmetic:

```
  3+6+pi
[1] 12.14159
```

Notice that you only need type the expression, in this case $3 + 6 + \pi$, and hit return. R automatically recalls the value of π under the name `pi`. Upon pressing return the command is executed, which results in a line of output. R treats data as vectors and matrices, and usually marks the row and column numbers for you. In this case the output is a scalar (1×1 array) so the one element is labeled as `[1]` and the numerical result is 12.141 59. Note that the output is printed with only five decimal places, but the calculation is performed at much higher precision. The result is rounded only for display purposes.

Now, let's assign a value to a variable. To do this we use the '`<-`' symbol. Notice this is two characters long, comprising a 'less than' followed immediately by a 'hyphen'. (The 'equals' sign is generally reserved for a slightly different purpose, as we shall soon see.) For example

```
  r <- 200
  pi*r^2
[1] 125663.7
```

assigns the value 200 to the variable r. The value of πr^2 is then evaluated with the expression `pi * r^2` and the result is returned. Other arithmetical operations such as division and subtraction have the usual symbols.

Note that R is case sensitive, so the variable `r` is different from the variable `R`.

The expression `1:20` will produce a sequence of integers $1, 2, \ldots, 20$. If we wished to assign a vector $\mathbf{x}$ with these values we would use

```
  x <- 1:20
  x
[1]  1  2  3  4  5  6  7  8  9 10 11 12 13 14 15
    16 17 18 19 20
```

where the first line assigns the sequence of integers from 1 to 20 to the variable $\mathbf{x}$. The second line, simply the name of the variable, prints its contents to the screen. In this case we see the elements are the integers as expected. R has many commands for operating on vectors and matrices, which is an essential part of efficient data analysis. For example, to calculate the sum, mean and variance of the numbers stored in $\mathbf{x}$ we simply use

```
  sum(x)
[1] 210
  mean(x)
[1] 10.5
  var(x)
[1] 35
```

and if we multiply a scalar by a vector (or matrix), this is equivalent to applying the multiplication on an element-by-element basis:

```
  2*x
[1]  2  4  6  8 10 12 14 16 18 20 22 24 26 28 30
    32 34 36 38 40
```

We can operate on the vector and assign the result to a new variable. For example, to calculate $y = (x - \bar{x})^2$ we would use

```
y <- ( x - mean(x) )^2
```

and to make a plot of y against x we use

```
plot(x, y)
```

To make the plot a bit more fancy, we change can the style, label the axes, and increase the character size

```
plot(x, y, type="l", xlab="x label", ylab="y axis too",
      main="title here", cex=2)
```

Notice how the command takes several optional inputs that specify other aspects of the plot – this is the way most R commands work, with all the options bundled into a single line.

If you wish to recall a previous command you can do this in a number of ways. Use the up arrow key to scroll through the list of previously entered commands. Or use

```
history()
```

to display the previous few commands (the default is to show the last 25 lines).

A.3 Entering data

One may load data into R in a variety of different ways. For small datasets you may enter the data into variables by hand. For example

```
x <- c(1,2,3,4,5)
y <- c(5.2,5.6,5.3,4.8,1.2)
```

This creates two variables, called `x` and `y`, each of which contains five numbers (which are combined into one object with the `c()` command). R will automatically decide whether the objects should be vectors of integers or floating point numbers (i.e. with decimal points). You may save these data to a plain text file, to avoid having to retype them in future. This can be done by first combining them as columns of a data frame, and writing this as a text table:

```
dat <- data.frame(x,y)
write.table(dat, file="myfile.txt", row.names=FALSE)
```

The file is saved to your current working directory. Check this using `getwd()` or change it using `setwd()`. If using the GUI (graphical user interface) for R (e.g. in Windows) you may use the **File** | **Change dir** menu to change the working directory. You may also load data from a file on your hard disc (in the working directory). For example

```
data.table <- read.table("myfile.txt", header=TRUE)
```

This will load the contents of the plain text file into the object `data.table`. For more help using this see `?read.table`. If you cannot remember the name or location of the data file, use the `file.choose()` command like this:

```
data.table <- read.table(file.choose())
```

One can also load data from the internet using `read.table()`, for example

```
data.table <- read.table(file =
    "http://www.statsci.org/data/general/waves.txt",
    header=TRUE)
n <- length(data.table$Waves)
time <- (1:n)*0.15
plot(time, data.table$Waves, type="o", pch=16,
     ylab="Force", xlab="Time (s)")
```

The first line reads a file from the specified internet location (which contains measurements on the force on a cylinder suspended in a tank of water in steps of 0.15 s) and stores its contents in an object called `data.table`. By default `data` is a table, which in this case happens to have only one column, which is called `data$Waves`. The name of the column is stored in the first (header) line of the file (note the `header=TRUE` argument of the `read.table()` command). The next two lines find the length of the dataset and create a vector of equal length called `time` which increases in 0.15 steps. The last line creates a simple plot of `Waves` against `time`.

One may also interactively edit data that are already in the memory, using a spreadsheet-like interface, as follows:

```
x <- edit(data.table$Waves)
```

A.4 Quitting R

To leave R type

```
q()
```

It will then ask you whether you want to save the "current workspace image" (i.e. any data, variables and settings you have defined during the current session). This will be loaded automatically when you next start R. My advice is to leave without saving the workspace image, although saving it can be useful if you plan to leave R and return later, and want to pick up exactly where you left it.

If you accidently saved the workspace from a previous session, and now want to start afresh, there are two ways to do this. You can either delete the workspace file, usually called `.RData` and stored in the current working directory, or you can restart R and then forget everything in the current working memory using the command

```
rm(list = ls())
```

A.5 More mathematics

Now for some matrix manipulation. Let's generate a simple matrix

```
i <- 1:6
mat <- matrix(i,nrow=2)
print(mat)
```

This makes a matrix with six elements, shaping it to have two rows (`nrow=2`), and therefore three columns. Now let's find the mean along each row and column

```
rowMeans(mat)
colMeans(mat)
```

This can be done with the `apply()` command, which is more general than the above:

```
apply(mat,1,mean)
apply(mat,2,mean)
```

The `apply()` command takes three inputs. These are the name of the data array to be manipulated, which dimension to use (1 = row; 2 = column) and what function to apply (e.g. `mean()`).

For example, we can also calculate variances

```
apply(mat,1,var)
apply(mat,2,var)
```

We could have used any other legitimate function (e.g. `sum()`, `sd()` etc.) in `apply(array,col/row,func)`. Now, we may append the matrix by adding the row/column means as an additional column/row

```
mat <- rbind(mat,apply(mat,2,mean))
mat <- cbind(mat,apply(mat,1,mean))
print(mat)
```

The commands `rbind()` and `cbind()` can be used to add a row or column to the data array, respectively.

The data array might be made clearer by giving names to the rows and columns

```
colnames(mat) <- c(1:3,"mean")
rownames(mat) <- c(1:2,"mean")
print(mat)
```

If we want to find the number of elements and the dimensions of the matrix

```
length(mat)
dim(mat)
```

A.6 Writing your own R scripts

Once you have started R you will be able to execute sequence valid R commands from the command-line by typing them in one by one. This is convenient for experimenting with commands (to make sure they do what you expect) and for performing very simple tasks. More complex or more repetitive tasks are better handled using scripts. A script is just a text file that contains a series of R commands in sequence. For example, we may write a script called `myscript.R`. To start a new script open a new file in a text editor, or use the built-in editor on Windows (click **File** | **New Script**, or to edit an existing script click **File** | **Open Script**). The script file might look like this:

```
# define the data array(s)
  d <- c(0.382, 0.949, 1.00, 0.532, 11.209, 9.449, 4.007,
         3.883)
  names <- c("Mercury", "Venus", "Earth", "Mars", "Jupiter",
             "Saturn", "Uranus", "Neptune")
# draw the dot chart
  dotchart(d[order(d)], labels=names[order(d)],
           xlim=c(0.0,12),
           xlab=expression(Diameter ~ (R[Earth])),
           lcolor="black")
```

In Windows you may then save the script (using the menu options **File** | **Save**). Try saving the above lines of code in a script called `myscript.R`. In order to execute the script (i.e. get R to perform the commands written in the file), we would use

```
source("myscript.R")
```

assuming the script is saved in the current working directory (see below). It is good practice to use R scripts to record any analysis you may wish to reproduce. Using scripts makes it easy to repeat, or modify then repeat, a piece of analysis. Including comments should make a script easier to understand for you and your colleagues. Any line starting with a hash sign # is a comment and R will ignore it.

You can display the location of the current working directory using the command

```
getwd()
```

This is the default location for reading and writing files. To change this use e.g.

```
setwd("Z:/My Documents/")
```

or use the **File** | **Change dir**... option in the Windows menu.

A.7 Producing graphics in R

R has a huge array of graphical capabilities. Here we shall concentrate on *base* graphics, which is the original graphics system for R and is available upon starting R. (Additional packages such as `lattice` and `ggplot2` add powerful new capabilities, but they are beyond the scope of this short introduction.)

There are two types of base graphics command in R: high level and low level. The high-level commands do several things: they open a graphics device (such as a window on the screen or a file on the hard disc), they define the size and shape of the graphic, the type of bounding box, the axis marks and labels, and finally plot some data. The most useful of the high-level plotting commands is `plot()`. Low-level graphics commands add data or detail to an already existing graphic (generated by a high-level graphics command). Let's see some examples.

```
x <- 1:100                # generate indices 1,2,...
y <- rnorm(length(x))     # generate random data
plot(x)                   # plot x-values in order
plot(y)                   # plot y-values in order
plot(x,y,type="l")        # plot y against x
```

The plot command usually has the form `plot(x,y,...)`, where `x` and `y` are the x and y values to be plotted and the additional arguments define the details of the plot. Let's make some more interesting data and make use of the options of the `plot()` command.

```
x <- seq(0,100)
y <- cumsum(rnorm(length(x))) + 50
plot(x,y,                 # (x,y) data to plot
  type="o",               # type of plot
  pch=16,                 # data symbol type
  lty=3,                  # line type
  cex=0.5,                # expansion factor
  xlim=c(0,100),          # x-axis limits
  ylim=c(30,70),          # y-axis limits
```

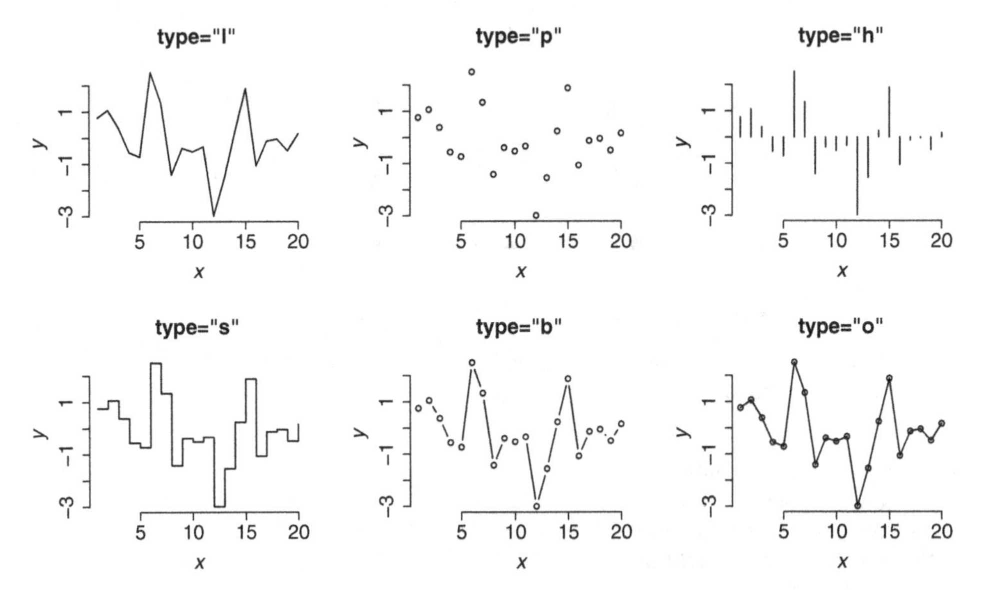

Figure A.1 Basic plot types. Each plot was generated with the same `plot()` command but with different values of the `type` argument.

```
   xlab="Time",                # x-axis label
   ylab="Value",               # y-axis label
   bty="n",                    # box type
   main="Title")               # title
```

If you want to plot on a logarithmic x-axis, y-axis or both, use the `log="x"`, `log="y"` or `log="xy"` argument, respectively. To see more options use the help function by typing

```
? plot                         # main help for plot()
? par                          # additional graphics parameters
```

The valid values for `type` are

- "`p`" for points
- "`l`" for lines
- "`b`" for both
- "`c`" for the lines part alone of "b"
- "`o`" for both overplotted
- "`h`" for "histogram" like (or "high-density") vertical lines
- "`s`" for stair steps
- "`S`" for other steps
- "`n`" for no plotting.

Several of these are illustrated in Figure A.1.

The different numbers for the plot symbols (`pch`) are shown in Figure A.2.

The different values for the type of box that surrounds the plotting region (`bty`) are

- "`o`" four-sided box
- "`l`" lower and left only

□ 0	○ 1	△ 2	+ 3	× 4
◇ 5	▽ 6	⊠ 7	✳ 8	⊕ 9
⊕ 10	✡ 11	⊞ 12	⊗ 13	⊠ 14
■ 15	● 16	▲ 17	◆ 18	● 19
• 20	○ 21	□ 22	◇ 23	△ 24

Figure A.2 The different symbols available from the `plot()` and `points()` plotting commands.

- "`7`" upper and right only
- "`c`" lower, upper and left only
- "`u`" lower, left and right only
- "`]`" upper, lower and right only
- "`n`" for no box.

Other high-level graphics commands include `pairs()` to produce a matrix of scatter plots; `coplot()` to produce a series of scatter plots separated by a third variable; `hist()` for histograms; `dotchart()` for dot charts; `image()` for intensity images; `contour()` for contour maps and `persp()` for projections of a surface.

We may also add to an existing plot using low-level graphics commands. These include `points()` to add new data points; `lines()` to add lines through data; `abline()` to add a straight line; `text()` to add text to a graphic; `legend()` to add a legend. For example, to add a horizontal line at $y = 50$,

```
abline(h=50,lty=2)
```

or we could calculate a theoretical curve and overlay it on the plot with the `lines()` command like this:

```
y <- 50 + 1*x - 0.02*x^2
lines(x,y)
```

A.8 Saving graphics in R

One of the most powerful features of R is its diverse range of graphics commands, as illustrated by the examples above (e.g. `demo(graphics)`). Producing graphics on screen is useful for interactively exploring and analysing data, but you will also need to produce and store or print copies of these graphics for your own notes or to illustrate reports and papers. Fortunately, R can export graphics in a range of formats. The way to achieve this is to open the relevant graphics *device*, then enter the commands to produce the graphics, then close the device. For example, to produce a JPEG output file:

```
jpeg(file="Routput.jpg",width = 480, height = 480)
plot(rnorm(50),type="l",ylab="y")
dev.off(dev.cur())
```

The first line prepares the output file (a JPEG file), gives it a name and sets the size (in pixels). The second line then generates the plot – notice that nothing happens on screen; the graphics are sent to the JPEG file. The last line tells R to finish with the JPEG file and restore the original graphics device (e.g. the screen). The output is a file called `Routput.jpg`. Note that between opening and closing the graphics device, you may use as many graphics commands as you wish.

One may produce output graphics in other formats including PDF, PNG, JPEG and PS. For a full list type `?device`).

Here is another example, showing how to produce a plot in PDF format:

```
i <- 0:30                           # create some fake x data
n <- dpois(i, lambda=12.8)          # create the fake y data
pdf(file = "Z:/My Documents/Routput.pdf")
par(oma=c(0, 1, 1, 1))              # set outer margin sizes
par(mar=c(5, 4, 4, 2)+0.1)          # set inner margin sizes
plot(i, n, type="p", bty="l", xlab="No. events",
     ylab="Frequency", lwd=5, pch=16, main="My plot",
     cex.lab=1.5, cex.axis=1.5)
lines(i, n, col="blue")             # join the dots
segments(i, 0, i, n)                # draw some vertical lines
dev.off(which = dev.cur())          # close the PDF file
```

The first two lines create some data to plot. The next line opens the PDF graphics device, gives the output file a name. By default the paper size is set to `A4` and text is written with the `Helvetica` font family. The next four lines set various graphics parameters (using the `par()` command). For the full range of parameters type `?par`. The first sets the outer margins of the page (`oma`; bottom, left, top and right); the second sets the inner margins (`mar`). Then the plotting command(s) are issued, including some further arguments such as `cex.lab=1.5` to increase the size of the axis labels. Finally the device is closed. This should produce a file called `Routput.pdf` which may be opened with any PDF viewer.

A.9 Good practice with R

Here is some general advice for using R.

- Read the documentation. The first place to start is the CRAN website, which hosts the official R documentation. The *Introduction to R* is available from http://cran.r-project.org/manuals.html.

  There is a range of online material kept here, and there is a growing literature of online and hardcopy books and courses using R.
- Use the built-in help documentation. When you are stuck with a specific command type e.g. `?plot`.
- Use scripts. Store batches of useful commands as script files. It is usually easier to modify existing scripts to perform slightly different tasks than write a new script from scratch.
- Arrange your scripts so they can be easily read by others. Use comment lines (beginning with a # symbol) throughout the script to explain the action of each part. Using appropriate names for variables and functions (e.g. `chisq.sim`) can make scripts easier to read.
- When using a new command try running it from the command line first, until you are sure you know how to use it. Once you are sure the command is doing what you require then add it to your script.

- It is almost always faster to operate on arrays (e.g. vectors, matrices) than it is to use a loop to apply an operation to each element of an array. Where possible use vector and matrix operations, and the `apply()` command on arrays in general.
- There are online communities using `R` to solve all manner of problems. Some resources worth looking at include Stack Overflow on programming (http://stackoverflow.com/questions/tagged/r), Cross Validated on statistics issues (http://stats.stackexchange.com/) and a digest of `R`-related blogs at R-bloggers (http://www.r-bloggers.com/).

Appendix B

Data case studies

This appendix discusses some of the datasets that are used as examples throughout the text. They are all available online from www.cambridge.org/9781107607590.

Although R can load files directly from the web, it is good practice to download the individual files to a local directory so that they can be used off-line. The files needed for this chapter are `rutherford.dat`, `reynolds.txt`, `hipparcos.txt.gz` and `pedroni.dat`.

B.1 Michelson's speed of light data

A. A. Michelson – known to students of physics for the famous Michelson–Morley experiment – made great advances in precision optical measurements, particularly the measurement of the speed of light. Here we shall use a set of 100 measurements of the speed of light in air taken in summer 1879 originally published by Michelson (1882) and reproduced by Stigler (1977).

The first few data values are shown in Table B.1. Each of the numbers represents the speed recorded in one 'run' of the apparatus, in units of km s^{-1}. Each run was in fact an average of several individual measurements. The 100 numbers, which together form a *sample*, are divided into five groups of 20, each group labelled an 'experiment'. The speed measurements are in principle continuous, but Michelson's data have been rounded to the nearest 10 km s^{-1}. Stigler (1977) applied Michelson's own corrections to the modern value of c to give a value of 299 734.5 km s^{-1} for the speed of light in air. This is the number Michelson was trying to measure in 1879.

R.Box B.1

Examine the speed-of-light data

The Michelson speed-of-light data come ready to use in R: no need to load them from an external file. Slightly erroneously the dataset is named `morley`. To view the full data simply type:

```
morley
```

which will produce 101 lines, one header line and 100 lines of data. The first column gives the row number; the other three columns are the data columns, labeled `Expt` for 'experiment', `Run` for the 'run' and `Speed` for the speed measured for that particular

Table B.1 *Subset of data from Michelson's 1879 speed-of-light experiment.*

Expt	Run	Speed −299 000 (km s^{-1})
1	1	850
1	2	740
1	3	900
1	4	1070
⋮	⋮	⋮

experiment–run combination. The data are in units of km s^{-1} after subtracting 299 000 km s^{-1}. To show only the speed data try

```
morley$Speed
plot(morley$Speed)
```

The following R.boxes give example code for presenting and analysing different aspects of these data: 2.1, 2.3, 2.4, 2.5, 2.6, 2.7, 2.8, 2.9, 2.10, 2.11, 2.16, 3.1, 3.2, 3.3 and 8.9.

B.2 Rutherford–Geiger radioactive decay

Rutherford and Geiger (1910) reported the results of an experiment into the rate of radioactive decay. For their experiment a small disc coated with polonium was placed in a vacuum tube, closed at one end with a zinc sulphide screen. The polonium decayed by release of alpha particles, which produced scintillations as they hit the screen. The scintillations were counted by eye with the aid of a microscope, and the counts registered on a paper tape. The tape was later divided into 2608 segments, each corresponding to a time interval of 1/8 minutes. The number of scintillations in each 7.5 s interval was tabulated; call this x_i with $i = 1, 2, \ldots, 2608$. The sequence of data started 3, 7, 4, 4, 2, 3, 2, 0, ...

Rutherford and Geiger's aim was to compare the observed distribution of the scintillation rate with a predicted distribution. The theory – described in a note by H. Bateman published along with the Rutherford and Geiger paper – assumed that individual radioactive decays were random and independent of each other, from which it can be shown that the scintillations per unit time should follow a Poisson distribution (see section 5.2.2)

$$p(x|\lambda) = \frac{\lambda^x e^{-\lambda}}{x!}. \tag{B.1}$$

The number of decays per unit time will not be exactly constant but fluctuate randomly between time intervals following the above distribution. This model has only one parameter, λ, the expected rate of decay. What mattered for their comparison of experiment with theory was not when each scintillation occurred, but the frequency distribution of scintillations per unit time. The data have been reduced from 2608 numbers (number of scintillations in each interval) to just 15 numbers, reproduced in Table B.2. The *frequency* indicates the number of intervals observed with a certain number of scintillations, e.g. there were 383

Table B.2 *Rutherford & Geiger's data.*

rate	0	1	2	3	4	5	6	7
freq	57	203	383	525	532	408	273	139
		8	9	10	11	12	13	14
		45	27	10	4	0	1	1

intervals showing exactly two counts. The data are discrete, being based on (integer) counts of individual particles.

R.Box B.2
Loading the Rutherford data

Download the data file `rutherford.dat` to your local disk and load the data into R using e.g.

```
rutherford <- read.table("rutherford.dat")
rutherford
```

The first line loads the entire data table into an object called `rutherford`. We can inspect its contents by simply typing its name at the command line. The columns in the data file are not labelled. By default R gives them names `V1` and `V2`. For clarity we create new variables `rate` and `freq` to hold these data.

```
rate <- rutherford$V1
freq <- rutherford$V2
```

R.Box B.3
A first look at the Rutherford data

The first thing we should do with data is make a plot

```
plot(rate, freq, type="h", bty="n",
     xlab="Rate (cts/interval)",
     ylab="Frequency", lwd=5, pch=16)
```

Using these data we can compute three important numbers: the total number of intervals recorded, the total number of scintillations counted and the mean rate of scintillations per interval.

```
n.obs <- sum(freq)
n.tot <- sum(freq * rate)
mean.rate <- n.tot / n.obs
```

The data record a total of $n_{\text{tot}} = 10\,097$ scintillations over $n_{\text{obs}} = 2608$ intervals, giving a mean rate of 3.87 counts/interval.

The following R.boxes give example code for presenting and analysing different aspects of these data: 2.2, 6.1, 7.3, 7.4, 7.5, 8.5, 8.6 and 8.7.

Table B.3 *Reynolds' data on fluid flow in a pipe, from Table 5 of Reynolds (1883).*

$\Delta P/\Delta L$ (m H_2O m^{-1})	v (m s^{-1})
0.000 80	0.0346
0.001 59	0.0646
0.002 39	0.0784
0.003 19	0.1262
0.003 98	0.1420
0.004 78	0.1711
0.005 58	0.1937
0.006 38	0.2260
0.007 17	0.2260
0.007 98	0.2455
0.008 77	0.2583
0.008 93	0.2583
0.009 57	0.2710
0.010 36	0.2774
0.011 17	0.2838
0.012 03	0.2905

B.3 A study of fluid flow

Reynolds (1883) described the results of several experiments to study the flow of fluids. These experiments, and the theories Reynolds published in 1895, were some of the most important in the development of the field of modern fluid mechanics.

The experiment involved running water through a long, thin cylindrical pipe under controlled conditions and recording how the rate of flow changes with the pressure. The pipe had a radius $R = 6.35 \times 10^{-3}$ m, and the temperature was approximately 8 °C. Reynolds' data from this experiment are reproduced in Table B.3. The velocity (in units of m s^{-1}) is the mean speed of the water and was obtained by measuring the volumetric flow rate Q (in units of m^3 s^{-1}) and dividing by the cross-sectional area of the pipe $v = Q/\pi R^2$. The pressure gradient $\Delta P/\Delta L$ was recorded in units of 'metres of water per meter'. We can convert to standard units by noting that a pressure of 1 m water is equivalent to 9.80665×10^3 Pa. (This comes from $P = \rho g$, where $\rho = 10^3$ kg m^{-3} is the density of water and $g = 9.80665$ N kg^{-1} is the standard gravity constant.)

R.Box B.4
Preparing Reynolds' data

Download the data file `reynolds.txt` to your local disk. The text file has a single line header that names the columns, so we can load the data into R using e.g.

```
reynolds <- read.table("reynolds.txt", header=TRUE)
```

Then we can convert the pressure gradient from Reynolds' units to units of Pa m^{-1} using the ppm (Pascals per meter of water) conversion factor and plot the data using

```
ppm <- 9.80665E3
dP <- reynolds$dP * ppm
v <- reynolds$v
plot(dP, v, bty="n", cex=1.5, pch=16,
     ylab="Velocity (m/s)",
     xlab="Pressure grad (Pa/m)",
     xlim=c(0,125), ylim=c(0,0.3))
```

We have used the `xlim` and `ylim` arguments to ensure the plot extends down to the true zero point of each axis. Now that we have Reynolds' data in our desired units we can store it as a data frame and save it to a file (e.g. `fluid.txt`). That way, when we need to examine these data we can load the data from this file and there is no need to repeat the unit conversions.

```
fluid <- data.frame(dP, v)
write.table(fluid, "fluid.txt",
            row.names=FALSE)
```

Assuming the flow is laminar (streamline, not turbulent) the pressure and the volumetric flow rate should be related by Poiseuille's equation

$$Q = v\pi R^2 = \frac{\pi R^4}{8\eta}\frac{\Delta P}{\Delta L}, \tag{B.2}$$

where η is the dynamic viscosity (units of Pa s). The mean velocity v should vary linearly with pressure gradient

$$v = \left(\frac{R^2}{8\eta}\right)\frac{\Delta P}{\Delta L}, \tag{B.3}$$

and the gradient of the line is given by $R^2/8\eta$.

The following R.boxes give example code for presenting and analysing different aspects of these data: 3.7, 3.8, 3.9, 6.2, 6.3, 6.4, 6.6, 6.7, 6.8, 6.9, 6.10, 7.1, 7.2, 7.7 and 7.8.

B.4 The HR diagram

The Hertzsprung–Russell (HR) diagram is one of the most important diagrams in all of astrophysics. It is essentially a scatter diagram showing, for a sample of stars, a measure of their luminosity (in magnitude units[1]) against a colour index (difference between magnitudes in different colour bands).

We can construct an HR diagram using the publically available catalogue produced using data from the European Space Agency's *Hipparcos* satellite (Perryman and ESA, 1997). The *Hipparcos* Main Catalogue contains 118 218 records (most of which correspond to stars), each with estimates of V (apparent, i.e. observed, magnitude in the V-band), $B - V$

[1] Magnitude is the logarithmic measure of brightness commonly used by astronomers. A smaller magnitude means a brighter star.

Table B.4 *Subset of data from the Hipparcos catalogue. The RA (right ascension, in hours, minutes and seconds) and declination (Dec, in degrees, minutes and seconds of arc) give the position on the sky. V is the apparent V-band magnitude (brightness in a green light filter), p is the parallax and $B - V$ is the colour index.*

HIP	RA (h m s)	Dec (d m s)	V (mag)	p (mas)	Error_p (mas)	$B - V$ (mag)	Error_{B-V} (mag)
2	0 0 0.91s	−19 29 55.8	9.27	21.90	3.10	0.999	0.002
3	0 0 1.20s	38 51 33.4	6.61	2.81	0.63	0.019	0.004
4	0 0 2.01s	−51 53 36.8	8.06	7.75	0.97	0.370	0.009
5	0 0 2.39s	−40 35 28.4	8.55	2.87	1.11	0.902	0.013
⋮	⋮	⋮	⋮	⋮	⋮	⋮	⋮

colour index (difference between magnitudes in B- and V-bands) and parallax (which can be used to estimate distance). Here we shall use a subset of the data with the best $B - V$ estimates, as stored in the data file `hipparcos.txt.gz`.[2] Table B.4 shows the first few records in the file. Unlike the other datasets this requires some additional processing to obtain the scientifically useful data.

R.Box B.5

Loading the Hipparcos data

With the file `hipparcos.txt.gz` stored in the current working directory, we can load it into R using the following commands:

```
hip <- read.table("hipparcos.txt.gz",
                  skip=53, header=FALSE,
                  na.string="-")
```

The first command loads the data file. But notice that we have explicitly stated `skip=53` and `na.string="-"`. The first of these forces the `read.table()` function to ignore the first 53 lines of the data file; these contain a description of the data table, not the actual data. The second tells `read.table()` that a hyphen symbol indicates missing data ("Not Available"). R functions often have special methods for handling missing data.

```
colnames(hip) <- c("HIP", "RA.h", "RA.m", "RA.s",
                  "DEC.d", "DEC.m", "DEC.s", "V",
                  "Plx", "Plx.err", "BV", "BV.err")
```

The data table does not include useful column names so we set these using the `colnames` command and a list of eight names (each of which is a character string).

[2] This was retrieved from the Vizier online service at http://vizier.u-strasbg.fr/viz-bin/VizieR on 17-Aug-2012. The catalogue `hip_main` was examined, and the records were filtered by accepting only sources for which the uncertainty on the colour index `e_B-V` was < 0.25 mag. This resulted in 85 509 matches, and from each of these 8 columns of data were stored in the file. The text file was then edited slightly in order to keep to a strict 8 column format, and mark empty columns, and compressed using `gzip`.

R.Box B.6
Removing the missing data

We can filter out any records with missing parallax data using the following line:

```
bad.data <- is.na(hip$Plx)
hip <- hip[!bad.data,]
```

Here, the `is.na()` function is used to identify all records where the parallax value is missing (has the special value `"NA"`), and we keep only those records that do not fall in this group. (The `!` symbol means a logical negation, NOT.) After this 85 446 records remain.

The data file contains eight columns of data, representing eight variables per source, including the apparent (observed) V magnitude, the $B - V$ colour index and the parallax p in units of milli-arcseconds. From the V_{obs} and p we can estimate the absolute magnitude V_{abs} (a logarithmic measure of luminosity). But first we must apply some additional filtering to the data to filter out any with poor parallax determination. A simple way to achieve this is to select only sources for which the relative (fractional) error on p is less than 5%.

R.Box B.7
Cleaning the data

In order to produce an HR diagram we need to remove the sources that are not stars, and for which we do not have a reasonable estimate of the distance from the source. We shall filter the data again to remove all sources without a good parallax estimate. We do this by selecting only those sources with a parallax uncertainly of $\leq 5\%$.

```
qual <- hip$Plx.err / abs(hip$Plx)
row.mask <- (qual < 0.05)
col.mask <- c(8, 9, 10, 11, 12)
hip.clean <- hip[row.mask, col.mask]
```

The first line computes the relative error on the parallax for each source (the `abs()` function is used because we are only interested in the absolute values, and a few anomalous records have negative parallaxes). The second line creates a vector whose elements are `TRUE` or `FALSE` depending on whether the $< 5\%$ error criterion was met. The third line selects only the last five columns of the data table. The last line selects the subset of rows (stars) and columns that meet the selection criterion. After this process 5740 records remain.

R.Box B.8
Computing the absolute magnitudes

We can compute, for each record, the distance in parsecs by converting the parallax to units of arcsec and taking its reciprocal. Using this we can compute the absolute V magnitude using equation B.4.

```
hip.clean$dist <- 1E3/hip.clean$Plx
```

```
hip.clean$V.abs <- hip.clean$V - 5 *
                   (log10(hip.clean$dist) - 1)
```

In the above lines we have effectively added some new columns called `dist` and `V.abs` to the `hip.clean` data array. We can now produce a simple HR diagram using the absolute V magnitudes and $B - V$ colour indices of the 'good' sources.

```
plot(hip.clean$BV, hip.clean$V.abs, cex=0.5,bty="n",
   xlim=c(-0.2, 2.0), ylim=c(15.5, -3),
   xlab="B-V (mag)", ylab="V.abs (mag)")
```

Note that the `ylim` parameter is used to specify the vertical axis runs from high to low, since lower absolute magnitude corresponds to higher luminosity.

R.Box B.9

Saving the cleaned, processed data

Now we have performed all this processing on the data – filtered the data to remove missing data points and excluded stars with poor parallax (distance) estimates, and computed the distances and absolute magnitudes – we can save the cleaned data in a file.

```
write.table(hip.clean, file="hip_clean.txt",
            row.names=FALSE)
```

Now when we need to analyse the data we can load the `hip_clean.txt` file rather than repeating the above data cleaning and processing. The new file is also a lot smaller than the original file (even when compressed).

For the surviving 5740 records with both good $B - V$ and p estimates, we compute the distance using $d = 1/p$ (if p is in units of arcseconds, then d is in units of parsecs), and then compute the absolute magnitude (i.e. distance-corrected magnitude, a logarithmic measure of the total luminosity) from

$$V_{\text{abs}} = V_{\text{obs}} - 5(\log_{10} d - 1). \tag{B.4}$$

We can now plot V_{abs} against $B - V$ to produce an HR diagram.

The following R.boxes give example code for presenting and analysing different aspects of these data: 2.12, 2.13 and 2.21.

B.5 A particle physics experiment

In 1952 Anderson and Fermi discovered a resonance in the pion–proton interaction at ~ 1.2 GeV. This was the first of many resonances discovered involving hadrons, which eventually led to the theory of quarks. Pedroni *et al.* (1978) described several experiments to investigate this interaction. One experiment was designed to measure the cross section for scattering of pions (π^+) on protons. The reaction is

$$\pi^+ + p \to \Delta^{++}(1232) \to \pi^+ + p.$$

Table B.5 *Pion cross-section data reproduced from Table 1 of Pedroni* et al. *(1978).*

E (MeV)	σ (mb)	Error$_\sigma$ (mb)	Length (cm)	Background (mb)
72.5	25.3	2.2	20	3.8
84.8	37.3	1.0	20	4.3
95.1	50.3	1.8	20	4.6
107.9	71.4	1.3	20	4.9
⋮	⋮	⋮	⋮	⋮
96.9	52.8	2.6	10	4.6
109.7	75.2	1.8	10	5.0
⋮	⋮	⋮	⋮	⋮

The $\Delta^{++}(1232)$ is a resonance – a short-lived particle – that in this case decays into the same kinds of particle that produced it.

The Pedroni *et al.* (1978) experiment used a beam of pions focused onto a sample of liquid hydrogen, behind which lay particle detectors. The kinetic energy of the incident pions (the energy in the beam) could be adjusted by the experimenters (this is the explanatory variable). The cross section (the response variable) was estimated by recording the change in the luminosity of the particle beam after passing through the sample, due to pion–nucleus interactions. This in itself required some substantial experimental work and data reduction.

The reduced data comprise a series of measurements of the scattering cross-section, σ, for $\pi^+ p$ interactions taken at different pion (kinetic) energies, E. The first few entries are shown in Table B.5. The kinetic energy E is in units of MeV (mega-electron Volts) and the cross-section σ is in units of mb (milli-barns). Along with each cross-section measurement is an error bar. The data were taken at two different settings for the length of the target (hydrogen container), and background levels were also estimated. For our purposes we shall combine the data for both settings of the target length, and ignore the data at $E > 313$ MeV, which have no background readings, as recommended by Pedroni *et al.* (1978).

R.Box B.10

Loading the pion scattering data

To load the data into R first download the file `pedroni.dat` and save it to your local disk, then use e.g.

```
dat <- read.table("pedroni.dat", header=TRUE)
mask <- (dat$energy <= 313)
dat <- dat[mask, ]
```

The second and third lines above remove any data at $E > 313$ MeV. The data are not supplied in order of ascending energy – because of the two different target length settings – so we first do a quick re-ordering and assign the results to some new variables. (We have no need for the `length` and `back` columns of the data array so we ignore them here.)

```
indx <- order(dat$energy)
x <- dat$energy[indx]
y <- dat$xsect[indx]
dy <- dat$error[indx]
```

R.Box B.11

Plotting the pion scattering data

To produce a basic plot with error bars use

```
plot(x, y, log="x", ylim=c(0, 210),
     pch=16, cex=1.0,
     xlab="Energy (MeV)",
     ylab=expression(sigma ~ (mb)) )
segments(x, y+dy, x, y-dy)
```

The model for the energy-dependent cross section of particle interactions is the Breit–Wigner (BW) model. For the sake of simplicity, we shall use a slightly simplified model here, based on the non-relativistic BW formula

$$\sigma(E) = N \frac{\Gamma^2/4}{(E - E_0)^2 + \Gamma^2/4}, \tag{B.5}$$

where $\sigma(E)$ is the energy-dependent cross section, N is a normalisation term, E_0 is the resonant energy of the interaction (measured in MeV) and Γ is the 'width' of the interaction (also measured in MeV). The width Γ is inversely related to the duration of the resonance, i.e. the lifetime of the Δ^{++} particle (by the famous uncertainty principle). In physics this curve is sometimes known as a Lorentzian distribution, and is often used to model the frequency response around a resonance. In statistics it is known as a Cauchy distribution.

The BW formula works well if the resonance is sharp, which means the width Γ is small compared with the energy E_0. This is not true for the Δ^{++} resonance, but we can get a reasonable approximation to the fuller theory by allowing the Γ factor to vary as a function of the energy as

$$\Gamma(E) = \Gamma_0 \left(\frac{E}{130\text{MeV}} \right)^{1/2}. \tag{B.6}$$

Here Γ_0 is the width at 130 MeV. Combining these two equations we have a model for $\sigma(E)$ with three unknown parameters: N, E_0 and Γ_0.

R.Box B.12

The Breit–Wigner model

The following code defines a new function called `model.pion()` that will compute the Breit–Wigner model with an energy-dependent width parameter

```
model.pion <- function(parm, x) {
  gam0 <- parm[1]  # width of resonance
```

```
    e0   <- parm[2]   # resonance energy
    norm <- parm[3]   # normalisation
    gam  <- gam0 *  (x/130)^(1/2)
    mod.y <- norm *  (gam^2/4)  / ((x - e0)^2 + gam^2/4)
    return(mod.y)
}
```

We can set the three parameters – resonance width, energy and the overall normalisation – to reasonable values and make a plot to see what this looks like as a function of energy:

```
parm.0 <- c(100, 180, 200)
mod.x <- seq(0, 400, by=1)
mod.y <- model.pion(parm.0, mod.x)
lines(mod.x, mod.y)           # to overlay the plot
plot(mod.x, mod.y, type="l") # make a new plot
```

In order to interpret the results of the experiment we also need to be aware of one more point. The energy data are given in terms of the π^+ kinetic energy in the reference frame of the laboratory. It is often useful to consider the centre-of-momentum reference frame (often called *cms*). The total energy in the cms is given by

$$E_{\text{cms}}^2 = m_1^2 + m_2^2 + 2m_2E_1 \tag{B.7}$$

(see e.g. Perkins, 2000, p. 6), where m_1 is the rest mass of the incident pion (with kinetic energy E in the laboratory frame), m_2 is the rest mass of the target proton (stationary in the laboratory frame) and E_1 is the total energy ($E + m_1$) of the pion in the laboratory frame. (Here we are using so-called natural units common in high-energy physics, in which $c = 1$ and the masses are in energy units, e.g. MeV, by $E = mc^2$.)

The following R.boxes give example code for presenting and analysing different aspects of these data: 6.5, 6.11, 6.8, 7.1, 7.2, 7.7 and 7.8, 8.8.

B.6 Atmospheric conditions in New York City

Bruntz *et al.* (1974) described a dataset comprising 111 measurements of four variables from a study of atmospheric conditions in New York City, May–September 1973. The variables are ozone concentration (in parts per billion), wind speed (in miles per hour, mph), temperature (in °F) and solar radiation (in Langleys). The data were used by Cleveland (1985, 1993) an example dataset for demonstrating visualisation of multivariate data.

R.Box B.13
The atmospheric data

The data do not need to be loaded from an external file, but come included in an add-on package called `lattice` that comes supplied with R. To initialise this package and prepare the data we run

```
data(environmental, package="lattice")
temp <- (environmental$temperature + 459.67) * 5/9
```

```
rad <- environmental$radiation * 41.84
wind <- environmental$wind * 0.44704
ozone <- environmental$ozone
env <- data.frame(ozone, rad, temp, wind)
```

The first line instructs R to load the dataset `environmental` that comes with the `lattice` package. This contains four variables: ozone concentration, solar radiation, temperature and wind speed. The other lines above convert the units of the variables (temperature from °F to K; solar radiation from Langleys to kJ m^{-2}; wind speed from mph to m s^{-2}). The last line produces a new 'data frame', a where each column is one of the transformed variables.

R.boxes 2.19 and 2.20 give code for generating different statistical plots using these data.

Appendix C

Combinations and permutations

In this chapter we present the basic ideas of *combinations* and *permutations*, sometimes known as combinatorial analysis.

C.1 Permutations

How many ways are there to re-arrange the letters {A, B, C}? By enumeration we find there are six: ABC, ACB, BAC, BCA, CAB, CBA. There are three possible choices for the first letter, two for the second and one for the third. There are thus $3 \times 2 \times 1$ arrangements. How many ways are there to re-arrange four letters, or 10, or N objects?

Let's consider a thought experiment in which N equal size sweets are placed in a bag. The sweets are each labeled consecutively from 1 to N. If we draw a sweet at random from the bag, it may have any of N possible labels. If we then draw another sweet from the bag (without replacing the first one), there are $N - 1$ possibilities for this second sweet. There are therefore $N(N-1)$ possible pairs of sweets that could be drawn. We can continue until the Nth draw, which will be of the only sweet left in the bag. The number of possible sequences of N sweets is therefore

$$N \times (N-1) \times (N-2) \times (N-3) \times \cdots \times 2 \times 1 = N! \tag{C.1}$$

We say the number of *permutations* is $N!$ (read N factorial), where a permutation is just an ordered arrangement of N objects or events.

Now, if we were to draw only four sweets starting from the full bag (with $N > 4$) there are

$$N \times (N-1) \times (N-2) \times (N-3) \tag{C.2}$$

possible sequences. This can be simplified by noting that

$$N \times (N-1) \times (N-2) \times (N-3) = \frac{N!}{(N-4)!}. \tag{C.3}$$

(You can check this by expanding the numerator and denominator of the right side of this equation and canceling the common terms.) More generally, the number of sequences of r sweets drawn from a bag containing N sweets is

$$N \times (N-1) \times (N-2) \cdots \times (N-r+1) = \frac{N!}{(N-r)!}. \tag{C.4}$$

We say there are $N!/(N-r)!$ permutations of r objects drawn from a set of N objects (with $N \geq r$). (In case you were wondering, by convention $0! = 1$.)

C.2 Combinations

A *combination* is an unordered sequence of objects or events. Returning to our bag, suppose we do not care about the order in which the sweets are drawn (e.g. we do not distinguish between 3, 17, 13, 11 or 3, 11, 13, 17), then we must account for this duplication. There are $r!$ ways to rearrange a sequence of r sweets (equation C.1), and we consider all these equivalent, so for every $r!$ permutation there is only one combination. Therefore, the number of combinations of length r drawn from N objects must be

$$\frac{N \times (N-1) \times (N-2) \cdots \times (N-r+1)}{r!} = \frac{N!}{(N-r)!r!}. \tag{C.5}$$

This expression has its own notation. The number of combinations is normally written in one of two different ways:

$$\frac{N!}{(N-r)!r!} \equiv \binom{N}{r} \equiv {}^{N}C_{r} \tag{C.6}$$

(read 'N choose r').

Box C.1
Examples of combinations

A panel is to be formed with three members drawn from a pool of 20. How many different panels are possible?

$$\binom{20}{3} = \frac{20!}{(20-3)!3!} = \frac{20 \times 19 \times 18}{3 \times 2 \times 1} = 1140.$$

How many possible hands of five cards can be dealt from a 52-card pack?

$$\binom{52}{5} = \frac{52!}{(52-5)!5!} = \frac{52 \times 51 \times 50 \times 49 \times 48}{5 \times 4 \times 3 \times 2 \times 1} = 2\,598\,960.$$

If each of these hands is equally probable (i.e. if the cards are well shuffled and there is no cheating), then the chance of getting the same hand twice in a row is $1/2\,598\,960 \approx 4 \times 10^{-7}$, less than one in a million.

R.Box C.1
Combinations in R

R has special functions to compute combinations. To calculate a factorial use simply

```
factorial(5)
```

So we can calculate $\binom{52}{5}$ using

```
factorial(52)/(factorial(52-5)*factorial(5))
```

But it is easier to use the special function

```
choose(52,5)
```

C.3 Probability of combinations

What about the probability of drawing particular sequences from the bag? The probability of drawing any particular numbered sweet from the bag is $1/N$. Once this sweet is withdrawn from the bag, the probability of drawing any of the remaining sweets from the bag then becomes $1/(N-1)$ since there are $N-1$ sweets remaining. By repeated application of the multiplication rule for independent events (equation 4.12), we find that the probability of any particular combination of sweets is simply the reciprocal of the number of combinations (equation C.5), since each is equally probable.

Appendix D

More on confidence intervals

Here we give more details of the construction of confidence intervals. As well as being more general than the treatment in Chapter 7, this should also help clarify the connection between hypothesis testing (section 7.4) and confidence intervals (section 7.6). These details are given in an appendix because they are not essential for following the text of Chapter 7. The argument given below follows that given in many other books (e.g. Barlow, 1989; Cowan, 1997; James, 2006), based on the original presentation of Jerzy Neyman.

Let's suppose we have obtained n data points of a response variable y_i, at different values of an explanatory variable x_i. We have a general model that predicts y as a function of x, but this depends on a parameter θ: $\mathrm{E}[y_i] = f(x_i, \theta)$. We can estimate the parameter θ using the data; for example, this might be the maximum likelihood estimate (MLE) $\hat{\theta}$.

Now, let's suppose we have two variations of the same model. One says that $\theta = \theta_0$, i.e. the parameter has some special value predicted in advance (this might be zero, for example). The other says that the parameter has some other value without specifying what it is, i.e. $\theta \neq \theta_0$. We'll call the first model H_0 and the second H_1, and it should be noticed that H_0 is a simple model and H_1 a composite model.

A hypothesis test between these two hypotheses involves comparing the observed value of the test statistic, $\hat{\theta}_{\mathrm{obs}}$, with the distribution of $\hat{\theta}$ under the assumption that H_0 is true, i.e. $p(\hat{\theta}|\theta = \theta_0)$. If the observed value is extreme, i.e. it lies in the tails of this distribution, then we reject H_0. Figure D.1 illustrates this idea with an α significance hypothesis test. This is a two-sided test, meaning that H_0 is rejected if $\hat{\theta}$ is too large or too small. What do we mean by too large or too small? We mean that, if H_0 is true, the probability of observing $\hat{\theta}$ below a low value a is $\alpha/2$, and the probability of observing $\hat{\theta}$ above a high value b is also $\alpha/2$. So the probability of observing $\hat{\theta}$ outside the interval (a, b) is α. We choose α to be small (in order to limit the number of type I errors). We reject H_0 if $\hat{\theta}_{\mathrm{obs}}$ falls outside this interval. We define the limits as

$$\Pr(\hat{\theta} \leq a|\theta = \theta_0) = \int_{-\infty}^{a} p(\hat{\theta}|\theta = \theta_0)\mathrm{d}\hat{\theta} = \alpha/2 \tag{D.1}$$

and

$$\Pr(\hat{\theta} \geq b|\theta = \theta_0) = \int_{b}^{+\infty} p(\hat{\theta}|\theta = \theta_0)\mathrm{d}\hat{\theta} = \alpha/2 \tag{D.2}$$

which can be compared to equation 7.4, except here the null hypothesis H_0 is $\theta = \theta_0$.

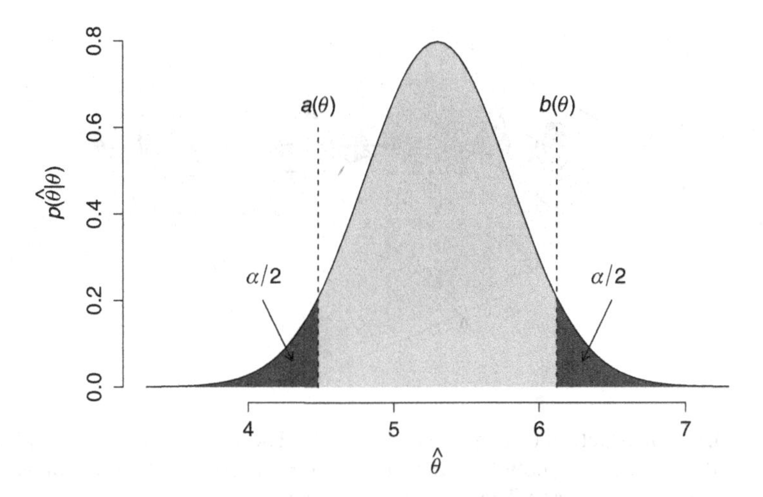

Figure D.1 Illustration of a two-sided hypothesis test. A statistic $\hat{\theta}$ has the distribution $p(\hat{\theta}|\theta = \theta_0)$ assuming some value $\theta = \theta_0$. A two-sided hypothesis test rejects the hypothesis $\theta = \theta_0$ when the observed value of $\hat{\theta}$ falls in the lower ($\hat{\theta} < a$) or upper ($\hat{\theta} > b$) tail of the distribution, which together contain α probability content.

From these two it should be easy to see that

$$\Pr(a < \hat{\theta} < b|\theta = \theta_0) = \int_a^b p(\hat{\theta}|\theta = \theta_0)\mathrm{d}\hat{\theta} = 1 - \alpha. \tag{D.3}$$

This means that, if H_0 is true (i.e. $\theta = \theta_0$), then 'in the long run' $\hat{\theta}$ will fall in the interval (a, b) in $1 - \alpha$ of repeats. This interval contains $1 - \alpha$ of the probability content of the distribution $p(\hat{\theta}|\theta = \theta_0)$. Our hypothesis test of $\theta = \theta_0$, using the $a < \hat{\theta} < b$ criterion, gives a type I error with probability α.

We can now return to the issue of confidence intervals. In this case we do not know the value of θ so we construct the interval for different values of θ, i.e. a and b as functions of θ.

$$\Pr(\hat{\theta} \leq a(\theta)) = \int_{-\infty}^{a} p(\hat{\theta}|\theta)\mathrm{d}\hat{\theta} = \alpha/2$$

$$\Pr(\hat{\theta} \geq b(\theta)) = \int_{b}^{+\infty} p(\hat{\theta}|\theta)\mathrm{d}\hat{\theta} = \alpha/2. \tag{D.4}$$

The functions $a(\theta)$ and $b(\theta)$ are shown as the smooth curves in Figure D.2. The region between these two curves is known as the *confidence band*. There is a $1 - \alpha$ probability that $\hat{\theta}$ will occur within the confidence band,

$$\Pr(a(\theta) < \hat{\theta} < b(\theta)) = 1 - \alpha. \tag{D.5}$$

This will be true for any values of θ, including the true value.

If the functions $a(\theta)$ and $b(\theta)$ increase monotonically with θ, then we can in principle find the inverse functions

$$A(\hat{\theta}) = a^{-1}(\hat{\theta}) \quad \text{and} \quad B(\hat{\theta}) = b^{-1}(\hat{\theta}). \tag{D.6}$$

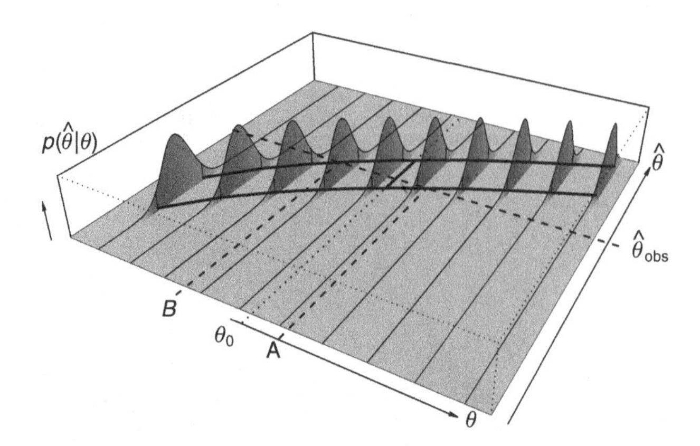

Figure D.2 Construction of confidence intervals. The interval (a, b) from Figure D.1 is calculated as a function of the unknown θ. (Figure D.1 corresponds to a single slice of this figure.) The curves of $a(\theta)$ and $b(\theta)$ are shown as the lower and upper smooth curves, respectively. The region between these two curves is the confidence band. The observed value of the statistic, $\hat{\theta}_{\text{obs}}$, is marked 'horizontally' on the plot, and the points where this crosses the confidence band (B and A) give the confidence interval (which is read off 'horizontally').

The inequalities $\hat{\theta} \le a(\theta)$ and $\hat{\theta} \ge b(\theta)$ correspond to

$$A(\hat{\theta}) \le \theta \quad \text{and} \quad B(\hat{\theta}) \ge \theta. \tag{D.7}$$

Equations D.4 can now be written as

$$\Pr(A(\hat{\theta}) \le \theta) = \alpha/2 \quad \text{and} \quad \Pr(B(\hat{\theta}) \ge \theta) = \alpha/2, \tag{D.8}$$

and combined to give

$$\Pr(B(\hat{\theta}) < \theta < A(\hat{\theta})) = 1 - \alpha. \tag{D.9}$$

The confidence interval on θ is given by (B, A) evaluated using the observed value $\hat{\theta}_{\text{obs}}$. The above equation holds irrespective of the (unknown) value of θ, and so this interval should include (cover) the true value with probability $1 - \alpha$.

Let's work through this argument graphically using Figure D.2. The confidence band is constructed 'vertically' as intervals on $\hat{\theta}$ as a function of θ. The functions $a(\theta)$ and $b(\theta)$ that define the accept/reject regions for a two-sided test of the hypothesis $\theta = \theta_0$ are calculated as a function of θ and marked as the lower and upper smooth curves, respectively. The region between these curves is the confidence band. The confidence band is the set of 'vertical' intervals (on $\hat{\theta}$), but we use it 'horizontally' (on θ) by marking the observed value $\hat{\theta}_{\text{obs}}$ as a horizontal line, finding where the curves $a(\theta)$ and $b(\theta)$ cross this line, and reading off the corresponding values on the 'horizontal' axis. Call these A and B (where $a(A) = \hat{\theta}_{\text{obs}}$ and $b(B) = \hat{\theta}_{\text{obs}}$). The data are random, and therefore so is $\hat{\theta}_{\text{obs}}$, which means the interval (B, A) is random and changes as the data change. The true value of the parameter θ_0 is not random: it has some specific value which is unknown. Now the question to ask is what is the probability that an interval (B, A), if constructed like this, includes the true value θ_0? (This is what we worked out above.)

$\hat{\theta}_{\text{obs}}$ is random, and crosses the ('vertical') confidence band for θ_0 with a probability of $1-\alpha$. Only when this happens does the ('horizontal') confidence interval (B, A) include the true value θ_0. Therefore, the confidence interval includes ('covers') the true value with a probability $1-\alpha$ (which means it does not include the true value with probability α). This is true whatever the true value of θ_0. Therefore, an interval so constructed should cover the true value with probability $1-\alpha$.

The graphical argument is actually quite simple and elegant, although it is quite easy to get confused between reading the 'horizontal' and 'vertical' axes of the θ–$\hat{\theta}$ diagram. The connection between a (two-sided) hypothesis test and a confidence interval should now be more clear. A two-sided hypothesis test with a significance level of α will reject the hypothesis $H_0 : \theta = \theta_0$ exactly when the $1-\alpha$ confidence interval on θ does not include the value θ_0. The $1-\alpha$ confidence interval can be thought of as the points around the estimate $\hat{\theta}$ at which you would not reject the hypothesis $\theta = \hat{\theta}$ with a significance level of α given the observation $\hat{\theta}_{\text{obs}}$.

The above argument gives the 'equal area' confidence interval, constructed using $\Pr(A(\hat{\theta}) \leq \theta) = \Pr(B(\hat{\theta}) \geq \theta) = \alpha/2$, but there is no need to restrict the intervals to equal upper and lower probability content. In some situations it is useful to place one-sided limits (upper/lower limits) on a parameter, using e.g. $\Pr(A(\hat{\theta}) \leq \theta) = \alpha$ only, which corresponds to a one-sided hypothesis test.

Miller and Miller (2003), Casella and Berger (2001) and Cowan (1997) give examples of the exact confidence interval formulae for simple estimators such as the sample mean with a normal or Poisson distribution, and the variance of a normal sample. But in practice this method of construction can be quite difficult. Often one will use instead approximation methods (e.g. based on the curvature of the log likelihood function; Chapter 7) or Monte Carlo methods (Chapter 8) to give approximate confidence intervals.

Appendix E

Glossary

addition rule The rule for combining probabilities of events $\Pr(A \cup B) = \Pr(A) + \Pr(B) - \Pr(A \cap B)$

Bayes' theorem In its simplest form Bayes' theorem states $\Pr(A|B) = \Pr(B|A)\Pr(A)/\Pr(B)$ and can be used to transpose the conditionals, i.e. relate $A|B$ to $B|A$

bias Quantifies how far the average statistic lies from the parameter it is estimating

binomial The distribution of the number of successes obtained in a series of n independent trials, when the probability of success (per trial) is p

bootstrap A resampling scheme allowing randomised datasets to be produced by resampling from the existing data

box plot Graphical representation of numerical data based on Tukey's five-number summary

bubble plot A type of scatter plot using circles as the plotting symbols, with the areas of the circles representing a third variable

central limit theorem States that the mean of n random variables tends to variable with a Gaussian distribution as $n \to \infty$ (under quite general conditions)

central moment The Nth-order central moment is the mean (expectation) of the Nth power of the deviation of a random variable from its mean

chi-square In general the sum of squares of ν independent normal variables has a chi-square distribution with ν degrees of freedom

chi-square statistic Test statistic formed from the (weighted) sum of squares of differences between data and prediction (model)

chi-square test A goodness-of-fit test that compares the chi-square statistic to a chi-square reference distribution

combination An unordered set of x objects from a set of n different objects (with $n \geq x$). The number of different combinations is given by ${}^nC_x = n!/((n-x)!x!)$

conditional probability Probability function that is conditional on some other statement being true. The conditional probability for event A given (conditional on) event B is written $\Pr(A|B)$

confidence interval An interval that has some stated probability of containing the value of some unknown population parameter (true value)

confidence level See coverage

confidence limits End points of a confidence interval

consistent (estimator) A consistent estimator converges on the expected value as the number of data points increases

correlation The tendency of one variable to increase or decrease together with another variable

correlation coefficient The covariance of two variables, normalised by the product of their standard deviations
covariance For paired random variables, the expected value of the product of the deviations from their means
coverage The probability with which a confidence interval for a parameter 'covers' the true (population) value of the parameter.
credible interval Bayesian counterpart of the frequentist confidence interval
cumulative distribution (cdf) Probability that the random variable X is less than or equal to x, for every value x: $F(x) = \Pr(X \leq x)$. Integral of the probability density
data Information, usually of a quantitative nature
deductive reasoning Reasoning from the general to the specific, or from causes to effects
degrees of freedom The number of values that can vary freely in a statistical calculation
density See **probability density**
distribution Description of the probability with which different values of a variable occur
distribution function Usually refers to the cumulative distribution function
error Difference between an observation or approximation and the true value (see also type I/II error)
estimate (n) The particular value of an estimator that is obtained from a particular sample of data and used to indicate the value of a parameter
estimate (v) The process of using data to make predictions about unknown model/population parameters
estimator Any quantity calculated from the sample data that is used to give information about an unknown quantity in the population. For example, the sample mean $\bar{x}$ is an estimator of the population (true) mean μ
event A subset of the possible outcomes of an experiment
expected value Mean value of a random variable, denoted $\mathrm{E}(X)$
explanatory variable A variable that is deliberately manipulated during an experiment, or selected during an observation (sometimes known as the independent variable). Usually plotted on the x axis in a two-dimensional (e.g. scatter) plot
Fisher information The variance of the score function. Can be estimated using the Hessian of the log likelihood
five-number summary A summary of a dataset or distribution comprising the least value, the lower quartile, the median, the upper quartile and the greatest value
Gaussian distribution Another name for the *normal* distribution
goodness-of-fit The agreement between data and hypothetical data predicted by some model
Hessian matrix A square matrix of all the second-order partial derivatives of a function
histogram A diagram using rectangles to represent frequency. Similar to a bar chart, except that a histogram may have unequal width bars, and it is the area of a bar that is proportional to the frequency it represents
hypothesis An assertion about the parameters or form of a model or population
hypothesis test A procedure for comparing hypotheses
iid shorthand for independent and identically distributed
independence Two events are said to be independent if the probability of one occurring is the same whether or not the other occurs, e.g. $\Pr(A|B) = \Pr(A)$
inductive reasoning Reasoning from the specific to the general, or from effects to causes
inference The process of arriving at a conclusion based on the evidence
interquartile range (IQR) the range between lower and upper quartiles (25th and 75th percentiles), which therefore encloses 50% of the values (around the median)
interval A set containing all numbers between two limits

joint probability The probability (density) associated with the possible values of a set of random variables. For two random variables X and Y the joint probability (density) function gives the probability for every possible pair of values the variables may take

likelihood Probability function for data conditional on some parameters, but considered as a function of the parameters (see sampling distribution)

likelihood ratio test A hypothesis test by comparing the ratio of (maximum) likelihoods for two models against a reference chi-square distribution

marginal probability The probability (density) function resulting from eliminating (by summation or integration) one or more variables from a joint distribution. The marginal distribution is in this way no longer dependent on the marginalized variable(s)

mean (of random variable) The mean, or expected value, of a discrete random variable is the sum of the products of the discrete values and their probabilities, $\mathrm{E}(X) = \sum_i p_i x_i$, and the mean of a continuous random variable is the analogous integral, $\mathrm{E}(X) = \int_{-\infty}^{+\infty} xp(x)\mathrm{d}x$. The mean value of a variable or population is usually denoted μ

mean (of sample) The arithmetic mean of a set of values is their sum divided by the number of values: $\hat{x} = 1/n \sum_{i=1}^{n} x_i$. The sample mean is an estimator of the expected value (the mean of the population)

median The 50th percentile. The median of a set of n values is the $(n+1)/2$th highest if n is odd, and the mean of the $n/2$ and $n/2+1$ highest values if n is even

mode The most frequent value in a set of data or a random variable. For a continuous random variable the modal value corresponds to the peak of its probability density function

model In general terms a model is simply an abstracted, and necessarily simplified, reflection of some part of reality. The terms 'hypothesis' and 'theory' are closely related and sometimes treated as synonyms

moment The Nth-order moment is the mean (expectation) of the Nth power of some variable (see also **central moment**)

Monte Carlo (methods) Methods in which sampling experiments, usually performed on a computer using pseudo-random numbers, are used to solve numerical problems

multiplication rule The rule for combining probabilities of events $\Pr(A \cap B) = \Pr(A|B)\Pr(B)$

multivariate A joint distribution involving two or more variables

noise Another name for stochastic processes

non-parametric A class of statistical methods that do not assume a specific distribution for the data (often known as distribution free)

normal A normal random variable is a continuous variable with pdf specified with two parameters, μ the mean and σ^2 the variance. The pdf is symmetric and has the classic 'bell-shaped curve' shape

null hypothesis In hypothesis testing the null hypothesis is the simpler hypothesis one accepts unless there is evidence to the contrary

***p*-value** Probability of getting a value of the test statistic as extreme as or more extreme than that observed, by chance, if the null hypothesis is true. Used to assess whether the data are surprising or not (assuming the null hypothesis)

parameter A value, usually unknown (and which therefore has to be estimated), used to represent a certain population/model characteristic

parametric A class of statistical methods that assume a model distribution for the data and proceed by estimating the parameters of the model based on the data

percentile Percentiles are the 99 values that divide the range of a variable into 100 intervals each with probability 0.01. The rth percentile is the value of a variable such than $r\%$ of the distribution lies below it

permutation The different possible orderings of a set of values. For example, the letters ABC can be arranged in six permutations: $ABC, ACB, BAC, BCA, CAB, CBA$. In general there are $n!$ permutations of n distinct objects. See also combination

point (estimate) A single value (point), such as the value of a parameter, estimated from data

Poisson distribution Probability function for a discrete variable X taking values $X = x$ with probability $\Pr(X = x) = (\lambda^x/x!)\exp\{-\lambda\}$ where λ is the mean (and variance)

population The set of all possible observations

posterior (probability) Represents the updated knowledge regarding the unknown model parameters after observing the data and other information pertaining to the unknown parameters

power (of test) Measures the ability of a hypothesis test to reject the null hypothesis when it is actually false (type II error)

precision The reciprocal of the variance

prior (probability) Quantifies knowledge regarding unknown quantities (e.g. model parameters) prior to observing the data. Used in Bayesian data analysis

probability A quantitative description of the likely occurrence of a particular event

probability density A probability density function (pdf) is a function that can be integrated to obtain the probability that the random variable takes a value in a given interval

proposition A well-formed statement that is either true or false

quantile The 0.3 quantile is the data point for which 30% of the data have lower values. See also **quartile** and **percentile**

quartile The quartiles divide the range of a variable (or dataset) into four intervals with equal probability (or frequency). The 25th, 50th and 75th percentiles are the lower quartile, median and upper quartile, respectively

random variable A variable that takes values in an unpredictable way

residual The difference between the data points observed and a model prediction

response variable A variable that is expected to change as a function of the explanatory variable(s), and is observed for this reason (sometimes called the dependent variable). Usually plotted on the y axis in a two-dimensional (e.g. scatter) plot

rms (root mean square) The positive square root of the variance (mean square deviation)

sample A set of observations (drawn from the population)

sample space The set of all possible outcomes of a random experiment

sampling distribution Describes probabilities associated with a statistic when a random sample is drawn from a population

scatter plot A simple plot in which pairs of data points are plotted as points on a Cartesian (x, y) graph

score The first partial derivative(s) of the log likelihood function

significance Measures the probability of a hypothesis test to wrongly reject the null hypothesis if it is in fact true (type I error)

size (of sample) The number of data points in the sample, usually labelled n

size (of a test) The probability of a type I error. See **type I error** and **hypothesis test**

standard deviation See **rms**

statistic A quantity calculated from a sample of data

statistical inference The use of information from a sample to draw conclusions (inferences) about the population from which the sample was taken

statistics The study of the collection and analysis of data
stochastic Another word for random
stochastic process A process whose outcome is not predictable
systematic error See **bias**
test statistic Quantity calculated from the data used as the basis of a **hypothesis test**
theory Usually a reasonably comprehensive mathematical framework for explaining some phenomena
time series A sequence of observations of a variable at different times
t **test** Student's *t* test is used to compare the mean of a sample of data with some predicted value or with that from another sample
type I error Null hypothesis is rejected when it is in fact true
type II error Null hypothesis is not rejected when it is in fact false
unbiased (estimator) An estimator without bias
variance Quantifies the spread or fluctuation of a variable in terms of the mean of the squares of the deviation between data values and the mean (or true) value
variate See random variable

Appendix F

Notation

Events and sets

A	an event
A^{C}	complement of A (the event 'not A')
$A \cup B$	the event 'A or B'
$A \cap B$	the event 'A and B'
Ω	the sample space (set of all events)
ω	an elementary event (member of Ω)
$\varnothing$	the null (empty) set

Probability notation

$\Pr(A)$	probability of A
$\Pr(A \cap B)$	joint probability of 'A and B'
$\Pr(A\|B)$	(conditional) probability of A, given B
X	random variable
x	particular value of random variable (as in $X = x$)
$p(x)$	probability density function for X as function of x
$p(x, y)$	joint probability density for x and y
$p(x\|y)$	conditional probability density for x, given y
$\mathrm{E}[X]$	expectation value of X
$\mathrm{V}[X]$	variance of X

Combinations

$n!$	n factorial $(= n(n-1)(n-2)\cdots(2)(1))$
${}^{n}C_x = \binom{n}{x}$	n choose x $(= \frac{n!}{x!(n-x)!})$

Models and summaries

$\mathbf{x} = \{x_1, \ldots, x_N\}$	vector of random variables
θ	parameter
$\boldsymbol{\theta} = \{\theta_1, \ldots, \theta_M\}$	vector of parameters
θ_0	(true) value of parameter
$\hat{\theta}$	estimator of parameter θ
$l(\theta)$	likelihood function
$L(\theta)$	log likelihood function $\ln[l(\theta)]$
H_0	null hypothesis

H_1	alternative hypothesis
T	test statistic
p	p-value (tail-area probability) from a significance test
α	significance level (of hypothesis test)
β	1 – power of hypothesis test
μ'_n	nth moment
μ_n	nth central moment
μ	true (population) mean
σ^2	true (population) variance
$\overline{x}$	sample mean
s_x^2	sample variance (for X)
s_{xy}	sample covariance (for X and Y)
r_{xy}	sample correlation coefficient
ρ	true (population) correlation coefficient
ν	degrees of freedom

References

Albert, J. (2007). *Bayesian Computation with R*. London: Springer.

Anscombe, F. J. (1973). Graphs in statistical analysis. *American Statistician*, **27**, 17–21.

Bailey, N. T. J. (1967). *The Mathematical Approach to Biology and Medicine*. London: Wiley.

Barlow, R. J. (1989). *Statistics: A Guide to the Use of Statistical Methods in the Physical Sciences (Manchester Physics Series)*, reprint edn. New York: Wiley–Blackwell.

Berger, J. O. and Berry, D. A. (1988). Statistical analysis and the illusion of objectivity. *American Scientist*, **76**, 159–165.

Bolstad, W. M. (2007). *Introduction to Bayesian Statistics*. Hoboken, NJ: Wiley.

Bruntz, S. M., Cleveland, W. S., Kleiner, B. and Warner, J. L. (1974). The dependence of ambient ozone on solar radiation, wind, temperature, and mixing height. In *Proceedings of the Symposium on Atmospheric Diffusion and Air Pollution*. Boston, MA: American Meteorological Society, pp. 125–128.

Campbell, L. and Garnett, W. (1882). *The Life of James Clerk Maxwell: With Selections from His Correspondence and Occasional Writings*. London: Macmillan.

Casella, G. and Berger, R. (2001). *Statistical Inference*, 2nd edn. Pacific Grove, CA: Duxbury Resource Center.

Charles, P. A. and Coe, M. J. (2006). *Optical, Ultraviolet and Infrared Observations of X-Ray Binaries*. Cambridge: Cambridge University Press, pp. 215–265.

Cleveland, W. S. (1985). *Elements of Graphing Data*. Monterey, CA: Wadsworth.

Cleveland, W. S. (1993). *Visualizing Data*. Summit, NJ: Hobart.

Cowan, G. (1997). *Statistical Data Analysis*. Oxford: Clarendon.

Efron, B. (2005). *Modern Science and the Bayesian–Frequentist Controversy*, Division of Biostatistics technical report, Stanford University.

Efron, B. and Tibshirani, R. J. (1993). *An Introduction to the Bootstrap*. New York: Chapman & Hall.

Fienberg, S. E. (2006). When did Bayesian inference become 'Bayesian'? *Bayesian Analysis*, **1**, 1–40.

Gardner, M. (1977). *Mathematical Carnival*. New York: Vintage.

Gelman, A., Carlin, J. B., Stern, H. S. and Rubin, D. B. (2003). *Bayesian Data Analysis*, 2nd edn. London: Chapman & Hall/CRC.

Gentle, J. E. (2003). *Random Number Generation and Monte Carlo Methods*. Statistics and Computing Series. New York: Springer.

Gregory, P. C. (2005). *Bayesian Logical Data Analysis for the Physical Sciences: A Comparative Approach with 'Mathematica' Support*. Cambridge: Cambridge University Press.

GUM. (2008). *Evaluation of Measurement Data – Guide to the Expression of Uncertainty in Measurement*. Joint Committee for Guides in Metrology (JCGM) technical report.

Howson, C. and Urback, P. (1991). Bayesian reasoning in science. *Nature*, **350**, 371–374.

James, F. (2006). *Statistical Methods in Experimental Physics*, 2nd edn. Singapore: World Scientific.

Jaynes, E. T. (2003). *Probability Theory: The Logic of Science*. Cambridge: Cambridge University Press.

Jeffreys, H. (1961). *Theory of Probability*, 2nd edn. Oxford: Oxford University Press.

Jeffreys, W. H. and Berger, J. O. (1992). Ockham's razor and Bayesian analysis. *American Scientist*, **80**, 64–72.

Lee, P. (2004). *Bayesian Statistics: An Introduction*, 3rd edn. London: Hodder.

Michelson, A. A. (1882). Experimental determination of the velocity of light: Made at the U.S. Naval Academy, Annapolis. [United States. Nautical Almanac Office.] *Astronomical Papers*, **1**, 109–145.

Miller, I. and Miller, M. (2003). *John E. Freund's Mathematical Statistics with Applications*. Upper Saddle River, NJ: Prentice Hall.

Pedroni, E., Gabathuler, K., Domingo, J. J. *et al.* (1978). A study of charge independence and symmetry from π^+ and π total cross sections on hydrogen and deuterium near the 3,3 resonance. *Nuclear Physics A*, **300**, 321–347.

Perkins, D. H. (2000). *Introduction to High Energy Physics*, 4th edn. Cambridge: Cambridge University Press.

Perryman, M. A. C. and ESA (eds). (1997). *The HIPPARCOS and TYCHO Catalogues. Astrometric and Photometric Star Catalogues Derived from the ESA HIPPARCOS Space Astrometry Mission*. ESA Special Publication, vol. 1200.

Reynolds, O. (1883). On the experimental investigation of the circumstances which determine whether the motion of water in parallel channels shall be direct or sinuous and of the law of resistance in parallel channels. *Philosophical Transactions of the Royal Society of London*, **174**, 935–982.

Russell, B. (1997). *The Problems of Philosophy*. New York: Oxford University Press.

Rutherford, E. and Geiger, H. (1910). The probability variations in the distribution of alpha particles. *Philosophical Magazine*, **20**, 698–707.

Sivia, D. S. and Skilling, J. (2006). *Data Analysis: A Bayesian Tutorial*, 2nd edn. Oxford: Oxford University Press.

Stigler, S. M. (1977). Do robust estimators work with real data? *The Annals of Statistics*, **5**(6), 1055–1098.

Tufte, E. R. (1986). *The Visual Display of Quantitative Information*. Cheshire, CT: Graphics.

Tukey, J. W. (1977). *Exploratory Data Analysis*. Reading, MA: Addison-Wesley.

Index

Printed in the United States
by Baker & Taylor Publisher Services